Voice of Life

Other books of interest

1. **भारतीय परम्परा में मृत्यु की अवधारणा** – शरीर का अन्तरण और दाहकर्म संस्कारः- लेखक ज्यान ज्युसेप्पे फिलिप्पी, अनुवादक बी०एल० धीमान, सम्पादक ज्यान्नी पेल्लेगीनी, ISBN 81-246-0313-8.
2. **Mṛtyu, Concept of Death in Indian Traditions** — Transformation of the Body and Funeral Rites; by G.G. Filippi; ISBN 81-246-0072-4.
3. **मृत्यु की दस्तक** – प्राचीन शास्त्र एवं आधुनिक ज्ञान, सम्पादकः- बैद्यनाथ सरस्वती, रामलखन मौर्य, ISBN 81-246-0314-6.
4. **Voice of Death** — Traditional Throught and Modern Science; edited by Baidyanath Saraswati; ISBN 81-246-0274-3.
5. **Yama** — The Glorious Lord of the Other World; by Kusum P. Merh; ISBN 81-246-0066-X.

Voice of Life

Traditional Thought and Modern Science

Edited by
Baidyanath Saraswati

D.K. Printworld (P) Ltd.
New Delhi

in association with

N.K. Bose Memorial Foundation
Varanasi

Cataloging in Publication Data — DK
[Courtesy: D.K. Agencies (P) Ltd. <docinfo@dkagencies.com>]

Voice of life : traditional thought and modern science /
edited by Baidyanath Saraswati.
vi, 151 p. 23 cm.
Contributed articles.
Includes index.
ISBN 8124603669

1. Life. 2. Philosophy and religion. 3. Science — Social aspects. I. Saraswati, Baidyanath, 1932–

DDC 113.8 22

ISBN 81-246-0366-9
First Published in India in 2006

Published by:
D.K. Printworld (P) Ltd.
Regd. office : *'Sri Kunj'*, F-52, Bali Nagar, New Delhi - 110 015
Phones : (011) 2545-3975; 2546-6019; *Fax* : (011) 2546-5926
E-mail : dkprintworld@vsnl.net *Web* : www.dkprintworld.com

in association with

N.K. Bose Memorial Foundation
B, 8/9, Bara Gambhir Singh, Varanasi - 221001 (UP)
Phone : (0542) 2275289; *E-mail* : nkbfoundation@yahoo.com

Printed by: D.K. Printworld (P) Ltd., New Delhi.

Contents

PART II
SCIENCE AND SOCIETY

Introduction

Baidyanath Saraswati

"From the viewpoint of anthropology, the acceptance of a theory depends greatly upon its fitness to holistic perspective." This is what Nirmal Kumar Bose used to say.

In my introduction to the first volume on *Traditional Thought and Modern Science*, I have shared with readers the *Voice of Death* published in 2005. The second volume is concerned with the *Voice of Life* at multiple levels, again through indologists, anthropologists, sociologists, philosophers, and physical and social scientists making an effort to re-language traditional thought in terms of modern science, making diverse interpretations at various levels.

Theme

In thinking out Life it is perfectly in order to compare the traditional vision with the modern science. Traditional thought is based on the world level. For the ordinary viewers there are two worlds: the Earth with the physical phenomena, and the Sky with the celestial phenomena. All existence is divided into two realms: the realm of the living and the realm of the dead. Both these realms are governed by Time.

To begin, it is important to raise questions. We see Life is a universal phenomenon; Death is also a universal phenomenon. But do they stand at the same level? Or is there a principle of Life on one hand, and a principle of Death on the other? Did they start off together, or sequentially, one after another? In whom do they continue? How do they unfold themselves? Do they belong to the same order of reality? Are they involved in maintaining the cosmic process? Is there a third entity beyond life and death? Can one find a key to this reality?

The world order is perpetuated by the Universal Feminine and the Universal Masculine. Time is the measure of all. The smallest unit of the cosmological time is the *yuga,* a creation age. The four *yuga*s make up a *mahāyuga*. One day of Brahmā, the creator, constitutes thousand fourfold *yuga*s. The Gods have incredibly long Life in comparison to the creatures of the Earth. Nevertheless, they also pass through the cycle of Life and Death.

How does Life manifest itself? The *Garbha Upaniṣad* describes the Life process thus: One night after the union, in a fertile period, arises a nodule (*kalil*); after seven nights it changes into bubble (*budbud*); in half a month arises the embryo (*piṇḍa*); after one month it becomes solid; when two months have passed the head appears; after three months the formation of the feet is completed; at four months happily originate the sex, the abdomen and the hips; at five months the back forms; after six months form the mouth, nose, eyes and ears; and after seven months the individual soul concludes the unification. At eight months all the parts are complete.

Is unborn Life significant? Yes, indeed, because it carries out the "occult effect" on one whose basic formation of personality is complete. The *Mahābhārata* story of Abhimanyu learning secrets of fighting a particular art of war while still in his mothers' womb is a case in point. That is to say, the genetically determined differences are formed in an "unborn" stage of Life.

What makes Life begin? The moment of birth marks the beginning of the early Life with the "inborn characters."

Parents contribute significantly in the formation of biological and psychological traits — the most fundamental properties of Life. The custom of endogamy is favoured to retain the ancestral genes. In astrology (*jyotiṣ*) the time of birth is a critical indicator of lifecycle, based on a purely mathematical calculation, giving minute details of events in one's Life, including the age at Death. Distinctions are made between the male and the female, because the qualities associated with the Universal Masculine and the Universal Feminine also give shape to Life.

How does lifecycle turn? Looking at the horoscope, an astrologer can predict the life-span of a person based on the astronomical clock. Further calculations of the time-scale will indicate the journey of Life hereafter, the future of the soul. Chronology of lifecycle is determined by the time at birth, as also the deeds of the person. The realm of the living-and-dying is the tangible world of Nature. Traditional cultures believe that the forces of death are directed toward the transmissions that limit the "Life's uncontrolled expansiveness." Putting the whole subject on a firm genetic code it is said that the essential pattern of male-and-female continues even after Death, as soon as the soul joins the "celestial Life." All the Gods in the sky have their feminine partners. Even the stars are identified on the basis of the male-female principle.

What lies beyond cause and effect, beyond Life and Death, beyond past and future? Indian cosmology holds that there are "powers" which generate Life, which indeed is the greatest. The "unborn Life" continues even after Death, because those powers knit the body. For the Hindus their ancestors are "ritually alive." They "feed" them, "invoke" them on ceremonial occasions, "treat" them as a part of their family and lineage. The world of ancestors (*pitṛloka*) and the world of Gods (*devaloka*) are invisible but communicable. The world of stars (including the Sun and the Moon) is visible. These worlds are not dead and gone. The human ancestors live in a unique realm of the "terminating transitions of Life." The soul is not dead, it is "re-born," and hence the "life" is eternal. There are three "real realms" of Life: the realm of the "unborn," the realm of the "living," and the realm of the "re-born" — each in a perpetual motion. The sages had said so. The cosmic cycle continues endlessly. Life-cycle follows the cosmic cycle. One of the several names of Śiva is *Mṛtyuñjaya,* the one who has conquered Death. In Greek, Roman and Indian mythology nectar (*amṛta*) is the drink that makes one everlasting. Ancients have been self-conscious of eternal LIFE, human and divine. The wheel of Life never stops. The concept of rebirth is associated with the body. The *Kaṭha Upaniṣad* says:-

> The seer is not born, and does not die,
> Springs not from any, nor becomes any;
> From everlasting he abides forever the ancient one,
> He does not perish, for it is the body that perishes.

The *Praśna Upaniṣad* has raised questions: What powers have knit the body? What powers give it Life? Which is the greatest? The sages have said that the powers are: air, fire, water, earth, speech, mind, light and hearing. All that has shape or no shape is Life.

Life is immortality. Life falls from Self as shadow falls from men. Life and Self are interwoven. Rising sun is the symbol of Life. God is greater than the greatest of all created. God and his sixteen phases are in the body. He creates Life; from Life, knowledge; from knowledge, air; from air, wind; from wind, light; from light, water; from water, earth; from earth, sense; from sense, mind; from mind, food; from food, vigour; from vigour, austerity; from austerity, revelation; from revelation, activities; from activities, world; from world, the names. When rivers mingle with the sea they lose their names and shapes and people speak of the sea only. So these sixteen phases, when they mingle with God, lose their names and shapes, and people speak of God only; man becomes the phaseless, the timeless.

The modern science of Life biology has also raised questions and put system into nature. However, its intellectual exercise is different from that of traditional thought. The evolutionary theory of *Homo Sapiens* presents a world-view called "ladder" mentality. The "ladder" symbolizes a climb through gradual stage: from simple to the complex, from instinct to intelligence, from non-living to living, from less conscious to more conscious, from inorganic elements to plants, from animals to humans, from disorder to order, to self-organization; in each case the world is planned to a ladder of progressive development. The ladder of evolution has been dropped recently in favour of different pathways: loops, punctuated equilibrium, non-linearity. However, the basic idea of ladder is maintained. Comparing the traditional thought with the priesthood of modern science, what do we hold in potential?

Voices

The voices of Life are loud. A Life-long mystery of beauty is there, but moments of forgetfulness, longing and belonging at the level of ego continue, according to one's nature and culture, undervalue or overvalue displayed. Fancy of imagination continues so long as the mind and consciousness remain. One clings to philosophy, religion, science and social life.

PHILOSOPHY AND RELIGION

The voicc of Life, raised in philosophy and religion, creates hierarchy and distinction between the people and the divinity, the secular and the sacred. Each person has a story, each culture has its own way of Life, each one runs through time. Everyone has to pass through a way of Life as prescribed in scriptures and social organization in the measure and manner of unique resources.

R.C. Sharma reflects on the Upaniṣadic concept of Life whose general system theory begins with *Brahman* or *Ātman*, which is conceived as the source of the entire Universe and Life. It refers to the beginning of the world from *Puruṣa* (Cosmic Man) from whose body the whole universe came up. The other sources of creation are generalized kind of way. For instance, the Soul (*Brahma* = Creator) evolved sky, then air, fire, water, earth, herbs, food, and lastly man. The issue of Life surfaces in various texts (*Kaṭhopaniṣad, Praśnopaniṣad, Muṇḍakopaniṣad, Bṛhadāraṇyaka, Śathapatha Brāhmaṇa, Chāndogya,* and other texts). The concept of Life as reflected in Upaniṣadic thought is of *jīvanamukta*, a stage where Life flows into salvation and where there is only a thin margin between Death and Life. Both are born simultaneously — Death in the form of hunger, and Life in the form of speech.

Uma Marina Vesci concentrates on the Vedic concept of Life, which is not something given from an alien source, but something out of its own inner strength, from *yajña* which is that of ensuring the emergence of Life out of inertia and its inner transformation. The aims of Life are *śanti* of the ever present Death.

B.D. Tripathi looks at life with covers that perceive transcendent principle, divine intelligence, and mysterious inner mechanism. Dealt with both pre-materialization and post-materialization — the subject remains a complicated one. Crucially, neither pre- nor post-materialism is easy enough to see differences. The unitary Life ceases to exist the moment its cover collapses. Everything is capable of transformation.

K.N. Sahay explores the mystery of Life, given the various levels: shades of mood, rhythm and environment, bearing with the threefold *guṇas*, according to the Āyurvedic system, perceived in Astrology, evolution and consciousness, purpose of Life and its destiny, horizontal man versus vertical man, spiritual Life, and models in the muddle, undoubtedly affecting the quality of human Life, almost frightening. This is in the air.

Physician Jogananda Jha begins with *Īśopaniṣad*, which says whatever exists in the world is the abode of *īś* — the ultimate power, the cosmic energy. So, Life cannot be anything except a part of that cosmic energy. This life-energy is perennial; it is imperishable. *Kaṭhopaniṣad* says: the seer is not born nor does he die.

Ajeya Jha, associated with technology, deals with the Web of Life, the sum total of all lives — all kinds of Life — symbiotic relationship that binds them, plus their vibrant interaction with the Earth, the Sun, the rains and the air. This web of life is often understood in terms of the wide variety of plants, animals and micro-organism. Extinction of Life has occurred as many as five times.

Contrasting with Ajeya Jha's ideas of *Web of Life*, based on physical law, Anthony Sudil, professor of Christian theology involved in inter-religious dialogue, draws out that the Christian perspective of Life is based on Bible and Biblical traditions. The Bible tells us that Life is sacred, and is created and gifted to humans and other living beings. As God is the author of Life, no one has the authority to misuse Life according to one's whims and fancies. Life has to be lived and used according to the specific plan and purpose of God which He has put into the created world. Life is used generally for a lifetime or duration of Life and specifically for an individual's way of Life. It is not a once-for-all historical event but as a type of super temporal behaviour existing alongside other possible types.

A.S. Inam Shastri, on the other hand, deals with religious concept of Life. According to Christian point of view, the humans are created by God, and hence have a responsibility to accept His reign. *Bhagavat-Gītā*, the most widely accepted Hindu scripture, says that the origin of the universe is not "out of nothing" but is a part of the Supreme God. The universe and all things in it are one, also sustained by Him through continued action. Life is highlighted in Holy *Quran*: "what is Life of this world, but play and amusements? The Life of this world is but empty. What is serious is the Life hereafter." As per the Islamic point of view, the Life of this world is just a struggle.

B.N. Saraswati sets forth: (a) the Manu principle which symbolizes the universal consciousness existing in the transcendent as also in the differentiated units of creation; (b) the Matter principle, the fundamental elements in their gross and subtle aspects of reality; (c) the Cosmic principle based on mathematical description of *Brahman* as zero; and (d) the Anthropological principle of Life grounded in the theory of evolution.

Indian tradition speaks of thought (or thinking about thought) as a transcendent phenomenon in the scheme of things. In developing the mathematical construction of *Brahman* (the ultimate) zero is infinite. Man's relationship with the Universe is explained by a powerful imagery of the cosmic Man, who is neither being nor non-being. From him the shining cosmic Egg was born, and from this again He himself emerged. In other words, He became the creator of his own creation. From this, eventually, the Four humans took birth. All kinds of Life emanated from the cosmic body, each invested with powers and functions.

PHYSICAL SCIENCE AND SOCIAL SCIENCE

We live in the age of machine. We have fantastic confidence — the triumph of materialism. Undoubtedly, the Newtonian view, with its doctrine of materialism, has contributed immensely to the advance of physical science. With new physics has developed new social science. It is fitting that the science that gave rise to materialism should also signal the demise of materialism; Quantum physics undermine materialism because it reveals that matter has far less "substance" than we might believe. Today, the ascendant nations and corporations are masters, not of land and material resources but of ideas and sophisticated technologies. In the overthrow of the old world-view, a paradigm shift in physical and social law is dramatically transforming human understanding of reality. Generally, the more science moves away from common sense, the harder it is to decide what constitutes a model and what is supposed to be a faithful description of the real world.

In the essays that follow, we offer the reader not only the physical science-view but also the thoughts of social scientists based on empirical studies, what may be called "touchstone of reality." Both the views have to be trusted.

Pratul Kumar Saraswati presents a brief history of Life, the evolution of man and technology standing firm on the ground of modern science, and looks into the patterns of evolution of Life during the past six hundred million years from simple to complex. However, brooding over the results, he affirms that any suggestion about the eternity of *Homo Sapiens* is wishful thinking. He returns to the history of Life, with uniqueness of extinction, and raises the question: Can we see the end to the terror of this history?

Surendra Nath Mishra raises fundamental questions: why does living matter differ so radically from non-living matter which also consists of several inanimate molecules? Why does the living organism appear to be more than much of the inanimate parts of its composition? What is life? What is its form and formula? He explains the biochemistry of living state, and energy transformation in living organism. However, he admits that it is still an unanswered question as to how the Life arises, or what may be the real meaning of Life?

Yolotl Gonvález Torres draws on Life, nature and the sacred in Mesoamerican cosmo vision, and takes an anthropological approach through which ritual cycles are performed, not only for humans but also for the growth of plants, especially maize. Earth is considered as the father and the mother of every living being. The Earth-Goddess is a feminine force in religious art of all times. Trees, by their size, their strength and their longevity stand out above all plants. Almost all the peoples have one or several trees, which they consider sacred, like the *pīpal* in India. Most of Aztec Gods which are anthropomorphic, also appear in the guise of animals. Animals, including snake, and other nature Gods such as Sun and eagles are iconographically represented. To open our eyes that nature is sacred, the only truth that these people can hope to possess is indeed a human one.

Onkar Prasad's theory of Life and Nature, as reflected in yet another ethnic group, named Santal, can be understood by analogy. Like the Mesoamerican cosmovision, the Santals are also tuned to the Earth, where plant sprang first and later the Karam tree and all vegetation grew. The bird perched upon the *sirom* plant and made a nest. There appeared two eggs from which two human beings, the primeval couple of the Santal, were born. Their powers of perception and language allow the integration of thought. According to them, these are the two kinds of human seed — the male and the female — both mobile and potent. The emergence of Life is described in brilliant clarity. The ultimate power to grant Life, however, rests with the *Bidhi Bidhānta*. These are aspects of *Ṭhākur Jīu*. Thus they alone mix the man-and-woman seeds. The event of child birth is essentially the same in tribal and non-tribal village communities. Though the seed is the perennial source of Life, it is the Supreme Being who alone grants Life.

A.S. Inam Shastri and Shaila Parveen provide an example of how modern science created a ghostly melee on the origin of Life on Earth. There are two views: (a) creation by God, and (b) natural evolution. The scientists' theory of Piltdown Man has come out as forgery. The essential point is that human limitations are inherent in the workings of Nature. It appears more credible *in principle*.

Umesh K. Singh moves towards understanding Life in sociological perspective, as presented by the 19th century authors such as Comte, Spencer and Durkheim. He tries to find touchstones of reality by going back to sociologists such as Marx, Hegel, Weber, and several other Western thinkers who take strong position rooted in modernity, many claiming that Life falls apart into unhallowed fragments.

Comparing traditional thought and modern science, it turns out that they are radically two different world-views. Traditional thought has to do with the wholeness of Life, taking into consideration the universal principle that in every man there is a spark of the divine. Modern thought, dominated by the materialistic world-view, has a compelling picture of the universe shaped by ideas, drawn from the realm of matter which seems to have no length at all.

Part I
Philosophy and Religion

1

Upaniṣadic Concept of Life

R.C. Sharma

THE ultimate aim of Upaniṣads is the realization of the Supreme Self; but in achieving this goal, one has to climb a number of steps, including the riddle of life and death. Life again has multiple facets, like the phenomenon of life, process of creation, philosophy of life, way of life or lifestyle, tradition, convention, norms and system of a specified period. Here, we are mainly concerned with the phenomenon and process of life, i.e. creation and Upaniṣadic view on the subject. Almost all the important Upaniṣads deal with this aspect briefly or elaborately.

Brahman or *Ātman* has been conceived as the source of the entire Universe and life emerges from It at Its will; and no paraphernalia is a pre-requisite for this process. The *Aitareya Upaniṣad* (associated with the *Ṛgveda*) says that in the beginning, even before the creation of the Universe, it was only the *Ātman*, nothing else. It desired to create the world[1] and it was created. First, the upper cosmos (*ambha*) was evolved, then sky, then this world (earth). . . .[2] It then desired to create the guardian deities (*lokapālas*), and for this purpose, it produced the Puruṣa

---

1. "ॐ आत्मा वा इदमेक एवाग्र आसीत्। नान्यत्किंचनमिषत्। स ईक्षत लोकान्नु सृजा इति।" *ऐतरेयोपनिषद्* 1.1.1.
2. स इमांल्लोकानसृजत। अम्भो मरीचीमरमापोऽदोम्भ। *ऐतरेयोपनिषद्* 1.1.2.

(Cosmic Man) from the water or It generated the Man (Brahmā) from the ocean. It then determined to evolve an Egg (*aṇḍam*) from which generated mouth (*mukha*) followed by speech, fire and two nostrils, which generated the smell from which evolved air or wind. Then were created two holes in which were set two eyes giving birth to sun (Āditya). The other two holes were provided with ears and from these sprang the directions. Then the skin was evolved and it created hair (*loma*), which gave birth to herbs and vegetables. Then heart was produced; from heart came out mind, which generated moon. Then navel was created, which generated the organ of ejection (*apāna*) and from it death was evolved. Then the organ of procreation (*śiśna*) appeared, which was the cause of semen, which produced water.[3]

Thus, the Body was created and the controlling deities of the faculties fell into the sea of this Great Body. The deities, who were subjected to hunger and thirst, requested the *Brahman* to provide a dwelling place for their installation (stay) where they could be fed[4]. For this purpose, the cow appeared and then horse, but they were not satisfied. The next stage was that of man (*puruṣa*) and the gods expressed their satisfaction to see the beautiful form (*sukṛtam*). To them, *Brahman* asked to enter the respective abodes.[5] Thus, the best body for living of gods of various faculties was shaped. Their functions, however, depended on hunger and thirst, which

---

3. तमभ्यत्तपतस्यामितप्तस्य मुखं निरभिपद्यत यथाण्डं मुखाद्वाग् वाचोऽग्निर्नासिके निरभि-द्येतां नासिकाभ्यां प्राणः प्राणाद्वायुरक्षिणी निरभिद्येतामक्षिभ्यां चक्षुश्चक्षुष आदित्यः कर्णौ निरभिद्येतां कर्णाभ्यां श्रोत्रं श्रोत्राद्दिशस्त्वङ् निरभिद्यत हृदयान्मनो मनसश्चन्द्रमा नाभि-र्निभिद्यत नाभ्यां अपानोऽपानान्मृत्युः शिश्नं निरभिद्यत शिश्नाद्रेतो रेतस आपः॥ *ऐतरेयोपनिषद्* 1.1.4.

4. ता एता देवताः सृष्टा अस्मिन्महत्यर्णवे प्रापतन् . . . ता एनमब्रुवन्नायतनं नः प्रजानीहि यस्मिन् प्रतिष्ठिता अन्नमदामेति। *ऐतरेयोपनिषद्* 1.2.1.

5. ताभ्यः पुरूषमानयना अब्रुवन् सुकृतं बतेति पुरुषो वाव सुकृतम्। ता अब्रवीद्यथायतनं प्रविशतेति। *ऐतरेयोपनिषद्* 1.2.3.

were asked to retain their share when oblation was to be offered to a deity.[6]

The *Taittirīya Upaniṣad* (associated with the *Taittirīya* branch of *Kṛṣṇayajurveda*) in the Brahmānanda-vallī states that from the Soul (*Brahman*) evolved sky, then air, fire, water, earth, herbs, food and lastly, the man.[7] The process begins only with the desire of *Brahman*.[8] Before it (creation) everything was unmanifested or concealed and the manifestation is from It only.[9]

In the *Kaṭhopaniṣad,* Yama, explaining the mystery of life to Naciketā observes that *Brahman* saw the First Born (*Hiraṇyagarbha-puruṣa*) and Aditi, the mother of gods was born.[10] The ruler or controller of everything resides in the heart of all and It measures like a thumb.[11] The breath, which is the basic element of life, is also controlled by It.[12] The birth is dependent on deeds. Sometimes, the soul enters the womb for getting a body and sometimes it is still.[13] As fire is seen in different forms in the world, similarly, the *Ātman* is seen reflected in different shapes.[14]

---

6 . . . तस्माद्यस्यै कस्यै च देवतायै हविर्गृह्यते भागिन्यावेवास्यामशनाया पिपासे भवतः॥ *ऐतरेयोपनिषद्* 1.2.5.

7. तस्माद्वा एतस्मादात्मन आकाशः सम्भूतः। आकाशाद्वायुः। वायोरग्निः। अग्नेरापः। अद्भ्यः पृथिवी। पृथिव्या ओषधयः ओषधीभ्योऽन्नम्। अन्नात्पुरूषः। . . . *तैत्तिरीयोपनिषद्*, वल्ली 2.1.

8. सोऽकामयत। बहुस्यां प्रजायेयेति . . . *तैत्तिरीयोपनिषद्* 2.6.

9. असद्वा इदमग्र आसीत् । ततो वैसदजायत । तदात्मानं स्वयमकुरुत। तस्मात्तत्सुकृतमुच्यत इति। *तैत्तिरीयोपनिषद्* 2.7.

10. *कठोपनिषद्* 2.1.6-7.

11. अङ्गुष्ठमात्रः पुरुषो मध्य आत्मनि तिष्ठति . . . *कठ* 2.1.12.

12. ऊर्ध्वं प्राणमुन्नयत्यपानं प्रत्यगस्यति . . . *कठ* 2.2.3.

13. *कठोपनिषद्* 2.2.7 (यथा कर्म तथाश्रुतम्)

14. अग्निर्यथैको भुवनं प्रविष्ठो रूपं रूपं प्रतिरूपो बभूव।
एकस्तथा सर्वभूतान्तरात्मा रूपं रूपं प्रतिरूपो बभूव॥ *कठ* 2.2.9.

In the *Praśnopaniṣad,* Kabandhī Kātyāyana requests Pippalāda sage to throw light on the creation[15] and the seer responds that the Lord of all (*prajāpati*) desired to have progeny and created the couple of food (*rayi*) and vital breath (*prāṇa*) with the idea that these will carry on the further process of creation.[16] The couple (*mithuna*) of *prāṇa* (breath) and *rayi* (food) is also conceived to be the sun and moon, i.e. male and female aspect.[17] The *rayi* is also considered to be *pitṛyāṇa* and sages who long for progeny follow this path, which is also known as *dakṣiṇamārga.*[18]

The opening verse of *Muṇḍakopaniṣad* informs that Brahmā was first to be born in the creation.[19] It further states that the *Brahman* acts as a spider in the creation of Universe. A spider releases the thread to spin the web and withdraws at its will, and the hair issues out of a living person, or the herbs grow automatically from the earth; similarly, the Universe emerges from *Akṣara,* i.e., *Brahman.*[20] The idea finds further elaboration in several hymns[21] and the essence is that everything evolves from *Brahman, Ātman, Akṣara* or *Puruṣa.*

---

15. *प्रश्नोपनिषद्* 1.1.3.
16. . . . प्रजाकामो वै प्रजापतिः . . . रयि च प्राणं चेत्येतौ मे बहुधा प्रजाः करिष्यत इति॥ *कठ* 1.1.4.
17. आदित्यो ह वै प्राणो रयिरेव चन्द्रमा रयिर्वा एतत् सर्व यन्मूर्त च तस्मान्मूर्तिरे रयिः॥ *प्रश्नो* 1.1.5.
18. ... ऋषयः प्रजाकामा दक्षिणं प्रतिपद्यन्ते एष ह वै रयिर्यः पितृपाणः॥ *प्रश्नो* 1.1.9.
19. ब्रह्मा देवानां प्रथमं सम्बभूव
    विश्वस्य कर्ता भुवनस्य गोप्ता। *मुण्डको* 1.1.
20. यथोर्णनाभिः सृजते गृह्णते च
    यथापृथ्व्यिामोषधयः सम्भवन्ति।
    यथा सतः पुरुषात्केशलोमानि
    तथाऽक्षरात् संभवतीह विश्वम्॥ *मुण्डको* 1.7.
21. *मुण्डको* 2.1.1-10.

The *Bṛhadāraṇyaka Upaniṣad* (associated with the *Śatapatha Brāhmaṇa*) is the biggest and foremost work in the pursuit of Self Realization. It is one of the earliest treatises on the subject. The concept of life has been discussed at different places, but the principle remains almost the same. Only *Ātman*, in the form of *Puruṣa*, was in existence even before the Universe. Being lonely, He became fearful; but the next moment He thought that if there is nothing else than Himself, then why and whom He should be afraid of.[22] This indicates that it was *jīvātmā* who felt fear and not the *Brahman*; and this is the secondary stage between *Brahman* and Brahmā or *Puruṣa*. He did not feel happy being alone and longed for a companion and transformed Himself into two, male and female combined, and this was the beginning of the couple, i.e., husband and wife; all men and women are the issues of that couple.[23]

The initiative taken by the first couple, explained as Manu and Śatarūpā in the commentary of Ādi Śaṅkarācārya, is rather curious and interesting. When the male started behaving amorously, the female part felt surprised and transformed herself into a cow and the male became a bull behaving amorously. The female then became a mare and the male became a horse. So, many forms of female and male in different species appeared; hence, the name Śatarūpā seems justified. The union of the first couple in multiple forms resulted in the progeny of countless creatures.[24] This was the beginning of the creation (*sṛṣṭi*).[25] The description gives vent to manifestation of desired, undesired, curious, amorous and erotic forms, which are seen reflected in literature and also the Indian art and architecture; and continued to be rendered or represented until late Medieval period.

---

22. बृहदारण्यको 1.4.2.
23. बृहदारण्यको 1.4.3.
24. बृहदारण्यको 1.4.4.
25. बृहदारण्यको 1.4.5.

The issue of life surfaces at other places, particularly in the delivery of sage Yājñavalkya to Janaka in the *Bṛhadāraṇyaka*. The life is associated in three parts or layers, the one where we live (*ihaloka*), the next world (*paraloka*) and the intermediary (*svapnasthāna*) experienced in dream and created by the "self."[26] The life is transmigrated from one body to the other like a leech (parasite) from one straw to other, or as a goldsmith takes apart a fragment of gold to fabricate another ornament.[27] This transmigration takes place when desires remain unfulfilled (*kāmāyamānaḥ*); but if nothing remains to be desired, at that stage the person (*akāmāyamānaḥ*) unites with the *Brahman*.[28]

The mystery of life has been unfolded at several places through the analogies in the *Chāndogya Upaniṣad*, a great work associated with the *Sāmaveda*. In the episode of Āruṇi and Śvetaketu, father and son respectively, the former says that in the beginning, only one existed and none second. It was *sat* which sometimes regarded as *asat* and from that non-existence issued existence.[29] That Existence or Non-existence contemplated to be in multi-forms. Consequently, fire or light was born and the fire also desired to be many and water was born. When a person suffers or sweats, it is because of fire and water generating from heat.[30] The waters, on their turn, also aspired to multiply and food was created; hence, the rain is the cause of multiplicity of food.[31]

The creatures have three categories, i.e. issued from eggs (*aṇḍaja*), from wombs (*jīvaja*) and from plants (*udbhijja*).[32] The

---

26. *बृहदारण्यको* 4.3.9.
27. *बृहदारण्यको* 4.4.3-4.
28. *बृहदारण्यको* 4.4.5-6.
29. सदेव सोम्येदमग्र आसीदेकमेवाद्वितीयम्। तद्धैक आहुरसदेवेदमग्र आसीदेकमेवाद्वितीयं तस्मादसतः सज्जात ॥ *छान्दोग्य* 6.2.1.
30. *छान्दोग्य* 6.2.3.
31. *छान्दोग्य* 6.2.4.
32. *छान्दोग्य* 6.3.1.

Deity (*Brahman*) further thought of entering into these three gods in the form of soul and this was the stage of manifestation in names and forms (*nāma-rūpa*).[33] Thus, life appeared in three-folds (*trivṛtam*), i.e., individual soul, name and form. Life flows in the body like juice in the tree. Whatever part of the tree is cut, the juice oozes out; if a part is separated from the main body, that part dries out and if the tree is cut from its roots, the entire tree goes dry and dies[34] as it is detached from the source of life (*ātmā*).

The seed of life is always subtle and remains beyond our perception. A large banyan tree sprouts from the smallest particle (atom) of seed, which cannot be noticed and this atom does not die even if the tree completely dries up.[35] Same is the case of individual soul, which leaves the body but does not die. It lives forever but dissolves or unites with *paramātman* or *Brahman* as salt with water.[36]

The Upaniṣadic philosophy of life is also rather unique. The people of that age did aim at opening the knots of life and death and realizing the Ultimate, i.e., *Brahman*. The conditions were congenial and the society on the whole was happy. Penance and austerity were practised but prosperity and pleasures were not neglected. The kings, like Janaka and Ajātaśatru, with all wealth at their disposal, remained busy in searching the Real Truth, i.e., *Ātma* or *Brahmajñāna*. On the other hand, the great sages, like Āruṇi and Yājñavalkya, possessing the perfect knowledge of *Brahman* and enjoying the Spiritual Bliss, were also the masters of big worldly fortunes. The prosperity, however, remained as means and not the aim. They were always prepared to part with it for

---

33. छान्दोग्य 6.3.2-3.
34. छान्दोग्य 6.11.1-2.
35. छान्दोग्य 6.12.1.
36. छान्दोग्य 6.12.2

higher values. Janaka was rightly called Videha, as despite being a great king, he often remained in trance. He had gifted the entire kingdom to Yājñavalkya (*Bṛhadāraṇyaka*) after listening from him the *Brahmavidyā*. Yājñavalkya on his turn would not accept any gift unless he was able to fully satisfy the king or his audience.

The ladies and housewives had also developed the same bent of mind as reflected from the episode of Maitreyī and Kātyāyanī (*Bṛhadāraṇyaka* II.4; V.1). Gārgī Vācaknavī (*Bṛhadāraṇyaka* III.6; III.8) was a brilliant debater and could challenge a stalwart like Yājñavalkya. This may, however, be remembered that some pitfalls were there and both sages and householders sometimes fell into them. The seeker of Truth, Satyakāma (*Chāndogya* IV.4) did not know about his father nor his mother knew, as she was serving many seers. But Satyakāma did realize the "knowledge." Śvetaketu (*Chāndogya* VI.1) and Bālāki (*Bṛhadāraṇyaka* II.1) falsely posed to be the possessor of *Ātmajñāna*, but later they had to admit their ignorance. The former learnt from his father and the latter from the king Ajātaśatru.

The story of Raikva and Janaśruti (*Chāndogya* IV.1) reveals that the attainment of Supreme Knowledge was not the monopoly of *Brahmarṣis* and *Rājarṣis*, but even a layperson could reach this goal by leading a good conduct and performing his duties with pure mind.

Thus, the concept of life as reflected in the Upaniṣads is of *Jīvanamukta*, a stage where life flows into salvation and where there is only a thin margin between death and life. Both are born simultaneously, death in the form of hunger and life in the form of speech. The death frightens the newly born baby to devour and it cries which is the creation of speech and death leaves the baby to live (*Bṛhadāraṇyaka* I.2.5). The play (*līlā*) of death and life goes on.

2

Vedic Concept of Life

Uma Marina Vesci

LIFE is a precious thing appreciated everywhere in the world. Life has a beginning, which can be witnessed at the birth of every child, and such a beginning, together with its subsequent growth, is so important that a "possible" coming to an end is rationally discarded — at least for oneself. The "I" who thinks and lives has no foreseeable end: such end (death) is an accident which occurs to "others" (even closest relations), and always by *mistake*, provoked by an external cause which could be avoided . . . . or not?

In all the religions and cultures of the world Life is a gift or a fact, and when Death imposes itself unavoidably on human consciousness, myths come on to *explain* why such an incongruity has appeared — to begin with. Man, all-over, deeply inside himself, refuses such a situation and, in one way or another, turns to a god, a hero, or a holyman, who can show the path towards immortality. In this way we can understand the importance of Life and the concern given to it. Yet, there is always a feeling of an End that could be *postponed* as much as possible, but never avoided altogether.

Thus, if Life is doomed to its End, how is it that the NK Bose Memorial Foundation Journal *Sacred Science Review*, has begun its journey with the theme on Death, reversing the order? The answer is quite simple, though a little strange for the non-Indian people: because here in India, right from the middle of Brāhmaṇic time, the order has been exactly that. In fact, even the questions are reversed: Why Life to begin with? How did Life come out and wherefrom? It is Life which has an incongruity, while a normal situation should be that of Rest, of Peace, of inertia. The 'normal' situation should be that of quietness, the emptiness, stretching out even before the separation of *sat* from *asat*. The Repose before any type of Activity is even conceived forth. Only if, at the beginning, one can count on a solid background of deep profundity, can activity — which means also Life — find a starting point on which bubbling energy spreads around in a multiplicity of concrete beings, *jīvana*.

Because this is Life. Not something "given" from an "alien source" (a god, for instance), but something *emerging* out of its own inner strength and having its cause from a primordial Cosmic Action : the *yajña*, at the beginning, performed by the gods over the Cosmic Man, Puruṣa (*Puruṣa Sūkta*, *R̥V* x. 90) and then acted directly by the main God, Prajāpati, over himself (*Sat*. *Br* III.I.I.).

Life is, therefore, an Activity, a Movement; is something emerging in its own rights out of a sticky background and has to be maintained, and kept going through the repetition of the same Act, or also to the performance of more common everyday works (on the body to ensure its growth and strength through food; in social context to ensure the means for acquiring such food). And this is another aspect of the correctness of giving such a sequential order of Life coming after Death. Not only Life emerges from the depth of Death, but it nourishes itself out of dead or dying (food), since Life is also present in grains, pulse, roots and vegetables, besides fruits (even vegetarian people and cows or other herbivores are not escaping killed matter).

But this reversed Indian position does not hint at a Life growing for ever in an increasing bubbling of energetic development. Indians are too wise to lean too long on such a possibility. Actually Life is not set in movement once for all in the performance of a *yajña,* even if the primordial one was performed by the Gods or by their Father and Lord "at the beginning." It is the product of the sacred Action, and such Action has to be repeated again and again to ensure its sustenance over and against the "solid background" of the peaceful Inertia. Surely the brāhmaṇas, both as theoretical philosophers and as practical performing priests, were potential *yajamāna*s to take over the burden of sponsoring such an important creative means. Besides promising them the gratification of their desires, like the birth of a son, the increase of wealth in the form of cattle or large crops in the fields, they also assure them to "build up" for them immortal bodies with which to enjoy a full after-life in heavens together with their gods. Thus, once those private advantages were taken care of, alongwith come also the other important features of the *yajña,* which is that of ensuring the emergence of life out of inertia and its inner transformation into a cooked, immortal and fixed matter.

This process is considered so effective and powerful that, when Prajāpati himself, the main sacrificial God, looks for means to solidify his body in the form of an altar (a special altar) he finds it in the vanishing boiling milk, the strangest and everlasting example of immortal transformation. On the same stake, the brāhmaṇas, all set up to increase the immortal power of their sacred Action, did encounter some other unexpected difficulties: to attain their aims (which, obviously, were not always granted) they had to enhance their sacrificial efficacy, which seemed to have developed heat during the procedure/process; strong heat becomes burning, that is to say: painful. The brāhmaṇas still found a cultic way to protect Heaven and Earth from such a suffering by isolating it in its vessel with a golden cover on the top and a silver plate at the bottom. Nevertheless, awareness of the pain involved in the process becomes significant. If Life was issued from such an action, that meant also that Life stemmed out from pain, an unavoidable pain. Thus, the beautiful Life for which all people yearned for revealed itself to be not only the laborious effort of emerging from the peaceful, balanced inertia, but also a painful one.

After such a discovery it cannot be said that Life was no longer appreciated. The majority of people, even at that far-away Brāhmaṇical time were still attracted to it in spite of such open recognition. What charmed however, and radically, was its philosophical interpretation. Life became a burden, a difficult effort grounded on suffering and therefore a situation that was better to be avoided altogether than to be cherished for. The only problem that now presented itself in full was the impossibility to avoid it or suppress it at will. Life not only was the product of a sacred action (even roughly performed) but was maintained with the concrete effects of the normal action performed by all living beings: their *karma.* This is no longer considered with a positive value and little by little people are taught by the emerging philosophy to get rid of its effects by slowly detaching themselves from their greediness and thereby now-disguised desires.

To sum it up, one can see that, apart from the Law of Karma which has to carry on and continue to push Life to drag along, already at the time of the early Upaniṣads, soon after the "discovery" of pain as part and parcel of sacrificial living Action, the philosophers and poets began to contest the issue. What is the use to force Life out of quietness again and again . . . and how long? Life is an effort, but growing old, this effort also looses its momentum and its attraction. If Life is an endless emergence from a vast restful background, why to take over oneself such an exhausting labour? Let quietness, peace, take over and revert again the *Aims of Life* into slow quiescence of the *Śānti* of the Everpresent Death.

3

Life with Covers

B.D. Tripathi

LIFE has two aspects, two dimensions, the *inner* and the *outer*. The *inner* is life *as such* and the *outer* is the indicator, cover, carrier, encompasser thereof. Human vision can go upto the cover and not beyond. Life being inner lies beyond human perception and comprehension. By way of inference, we may describe life as the inner content of cover, as piece of energy or force which initiates its holder to move from one stage of *being* to the other. Life may be viewed as a glowing piece which illumines its holder. Life and cover are mutually interdependent entities. Life cannot exist without cover and cover cannot exist as a living item without life. Life is a force, energy which Freud calls Eros and Jung calls Libido. Life has no physical attributes. Life is a mystery to which human vision has no access. Human mind cannot resolve the mystery and riddle of life as it lies far beyond his perceptual range. Life is there but it cannot be caught and deciphered. It is the cover of this energy which may be held and subjected to scientific scrutiny.

Now, let us define life. Life is a *mysterious inner mechanism* which initiates movement in its cover, body or holder. Till life-force exists in a particular unit-cover, it moves from one state of *being* to the other. The moment the glow of life extinguishes, the body or cover ceases to exist, ceases to be. Every unitary life-cover may be likened to a sentence. Like a sentence, it has a beginning and an end. In sharp contrast to unitary cover, there is universal cover of life. Universal cover remains full for ever, encompasses countless unitary covers within its fold. Universal cover is ever-lasting, eternal.

Unitary life-cover has two phases of being: pre-materialisation and post-materialisation. Pre-materialisation phase of life-cover is an underground reality. Even at this stage of reality, the cover feels the force of life; it moves and lives as initiated by life; it grows as initiated by life. In the course of time, underground reality moves to the state of ground reality. This is the post-materialisation phase of life-cover. Post-materialisation phase of cover may be seen as passing through three stages namely, growth, maintenance and decay. In the first stage, life-cover grows. The process of growth is self-initiated, though it is helped or restrained by the external factors. In the second or maintenance phase, the tendency of growth and decay remains equally matched. This is a plateau or stability stage of life-cover. The third and the last stage of life-cover is decay. At this stage, the process of decay outpaces the process of growth. The cover, therefore, gradually moves toward the end-point of its entity. The moment the process of decay completely overtakes the process of growth, the cover falls within the lap of death. The force which initiates the cover to move from its point of beginning to the point of fag end is life. Life is another name for movement, flow and blow. The absence of flow, blow and movement is death. Diagrammatically speaking, life-cover is like an inverted U on the axis of time (span of life).

From sustenance view-point, there is mutual interdependence between the force and the cover of life. On the one hand, force of life initiates the cover to move from one stage to the other till the end-point; on the other, the cover protects life from external exigencies as a palm does in respect of lamp flames.

Life is associated with covers which range from vegetation to man. In between these two extremes, there are several island-covers more evolved than vegetation and less evolved than man. Life's singular function, in respect of all sorts of covers, is to initiate them to move from one state or *being* to the other and leave them motionless at the end.

From the association between life and covers, there emerges the phenomenon of consciousness. The quantum of consciousness is least in the world of vegetation and most in the humans. In the mid-categories of covers, amount of consciousness is more than vegetation and less than man. The emergence of consciousness in the cover may be likened to the blooming of flower in plants.

In the covers of latter categories, there emerges mind from the phenomenon of consciousness through the mediation of brain. On instructions from life, mind weaves certain tendencies, propensities and pre-dispositions in the form of instinct which go to protect the interest of the covers, overlooking others' concern. Most of the instincts are self-protective, self-serving, self-enhancing and self-actualising irrespective of the interests of other unitary covers. In other words, the fruit that arises from the flower of consciousness is "Īd." Īd is self-confined, self-concerned and self-limited. Īd is the primary and raw-disposition of the animal cover. Later, culture evolves Ego and Super-Ego from the raw disposition called Īd. It is this Ego and Super-Ego which jointly strike a balance between the interest of society and the individual.

Life initiates movement in the cover and the cover reciprocates this gesture of life by providing protection and sustenance to the former. In the course of movement from one phase to the other, the cover undergoes wear and tear in the last phase. It gradually loses its capacity to protect life lying within its system. At the last stage, unprotected by the worn-out cover, life moves towards the point of extinction or dissipation and the unitary cover reaches its point of terminus. At this point, the entire structure of life along with its superstructure, consciousness and mind, collapses. No life, no cover; no cover, no consciousness; and finally no mind. Further birth of life in subsequent covers is more a fiction than a fact.

Shakespeare in his book *Macbeth* calls life "a walking shadow" and "a tale told by idiot, full of sound and fury, signifying nothing." But in my view, life is not a walking shadow nor is it an insignificant tale. In fact, life is the only author and director of the entire drama we come across the world.

Life has been eulogised and deified as *ātman, jīva, Rīha* or even God in different religious systems. The unitary life ceases to exist the moment its cover collapses. Unitary life is not eternal and everlasting as assumed in different Faiths. In my view, by calling unitary life eternal and everlasting, we heap such attributes on it which are simply not there.

4

Exploring Mystery of Life

K.N. Sahay

LIFE means "the condition which distinguishes active animals and plants from inorganic matter, including the capacity for growth, functional activity, and continual change preceding death" (Allen, 1990).

However, when we talk of human life, it does not have a vacuum. It has a bearing with society, and the society has a culture, its norms and traditions. Life is not a complete thing unless you refer to the "life-cycle" as such. It starts from birth of a child, adolescence, young age, married, adulthood, old age, and death. Hindus believe in rebirth, and so again, it starts from conception and birth. Human life is categorised into different sectors as said above, and hence various types of basic traditions, norms and behaviour.

Life has different dimensions — physical, environmental, social, cultural, religious, spiritual, psychological and the like. Sociologists like Radcliffe-Brown (1964) tell us that the social life and its structure which has different phenomena, all constitute a complex mechanism by which a social structure exists and persists.

Shades of Mood in Life

Man has different shades of mood in life, all depending on the circumstances. It may be cheerful, worried, enthusiastic, depressive, pragmatic, sentimental, philosophical, mystic, peaceful, merciful, indifferent or the like. A number of poems in different literature tried to depict such moods, and a few of them may be referred to here. The Hindi poet, Sumitra Nandan Pant in his book: *Gunjan* talks about the *Sukh-Dukh* (Joy and Sorrow) and tells that transition of the twilight and the break of dawn have tears as well as joy. He tells:

यह साँझ उषा का आँगन
आलिंगन विरह मिलन का
चिर हास अश्रु मय आनन
रे इस मानव जीवन का. . . .

Another poet, R.C. Prasad in his book: *Kalāpi* refers to *Jīvan Kā Jharnā* (Spring of Life) and tells that life is like a spring. Its water is just like a passion which is arbitrarily running fast in between the two banks of joy and sorrow. He tells:

यह जीवन क्या है निर्झर है
यह मस्ती उसका पानी है
सुख–दुख के दोनों तीरों में
चल रहा राह मनमानी है. . . .

Mahadevi Verma, in her poem *Virah Kā Jaljat Jīvan,* talks about the rainbow during rainy season which has seven colours presenting different moods. She tells:

"जीवन जल–कण से निर्मित सा
चाह इन्द्रधनु से रंजित सा
सजल मेघ सा धूमिल है
जग चिर नूतन सकरुण पुलकित सा. . . ."

The English poem, "We are such stuff" of William Shakespeare is philosophical and melancholic. He talks about the pageantry of life which fades.

Our revels now are ended. These our actors,
As I foretold you, were all spirits and
Are melted into air, into thin air:
And, like the baseless fabric of this vision,

The cloud-capped towers, the gorgeous palaces,
The solemn temples, the great globe itself,
Yea, all which it inherits, shall dissolve
And, like this insubstantial pageant faded,
Leave not a rack behind. We are such stuff
As dreams are made on, and our little life
Is rounded with a sleep.

In the other poem, "All the world's a stage," Shakespeare talks about the human life and its activities which constitute "plays of drama in life." He tells:

All the World's a stage.
And all the men and women merely players:
They have their exits and their entrances;
And one man in his time plays many parts,
His acts being seven ages. . . .

Henry Howard, Earl of Surrey, in his poem, "Happy Life" presents interesting criteria of happiness. He tells:

Martial, the things that do attain.
The happy life be these, I find:-
The richness left, not got with pain;
The fruitful ground, the quiet mind;

The equal friend; not grudge, not strife;
No charge of rule, nor governance;
Without disease, the healthful life,
The household of continuance;

The mean diet, no delicate fare;
True wisdom joined with simpleness;
The night discharged of all care,
Where wine the wit may not oppress;

The faithful wife, without debate;
Such sleeps as may beguile the night,
Contended with thine own estate,
Ne wish for death, ne fear his might.

Life: Rhythm and Environment

Environment generally constitutes land, water, air and living organism. It has three basic components: Physical Environment, Biological Environment and Social Environment. Social Environment includes interactions of social behaviour and norms among human beings, in a society, and they impinge upon the first two aspects.

In primevel society Man and Nature had perfect tuning and *rhythm*. Nature was looked at with awe and reverence. The Vedas, folklore, scriptures or other Sanskritic texts bearing with religion, speak of the need for harmony with the universe in which men, animals, birds, insects, vegetation, forest, hills, rivers and climate coexist, and have social and harmonious existence. "The ecological balance is inherent in the very process of creation. Everywhere, the seers, poets and thinkers through the ages, have referred to the need for living in harmony with environment" (Chaturvedi, 1989:ix).

The *Taittirīyopaniṣad* refers to the relationship between man and his environment in its totality and stresses complete harmony and its inter-dependence between them in order to attain real prosperity (Nikhilananda, 1949). It refers to the famous *mantra* (*śloka*) given below:

> ओ३म् द्यौः शान्तिरन्तरिक्ष॑ शान्तिः
> पृथ्वी शान्तिष्यः शान्तिरोषधयः शान्तिः ।
> वनस्पतयः शान्ति र्विश्वे देवाः शान्तिर्ब्रह्म शान्तिः
> सर्व॑शान्तिः शान्तिरेव शान्तिः सा मा शान्तिरेधि ॥
> ओ३म् शान्तिः ! शान्तिः !! शान्तिः !!!

Besides, in Indian art *bhangas* (adaptation of posture) were common in the second century CE and explored states of poise, counterpoise, and equilibrium. Life was expressed through *rasas*, flavours, atmosphere in peacefulness or tranquility. "In short, the values of life were based upon the appreciation of atmosphere, or flavours of one's environment. . . . These *rasas* reviewed the central cardinal precepts of life within the simple format of man's existence, and formed a natural ecological harmony with his surroundings (Imam, 1990: 267)."

Truly, man had traditionally availed of the benevolence of Nature, but in a modest and nearly harmless manner it had natural ecological harmony with the surroundings. However, with the onset of technological accomplishments people became arrogant about their prowess, and the requirements of the industrial civilization and population explosion gave way to an increasing exploitative relationship which disturbed the rhythm of the ecological harmony (Sahay, 1999: 306-8).

Inamdar (1989: 353) tells that "Environment, in biology, comprises the whole range of external influences acting on an organism, of both the physical and the biological forces of nature surrounding the human individual. "Environment has been defined as "the aggregate of all the external conditions and influences affecting the life and development of an organism." (Sahay, 307)

Human Life: Bearing with Triguṇas

The ancient concept of *Triguṇa* i.e., sāttvic, rājasic and tāmasic — the three basic attributes or innate disposition of man as envisaged in *Bhagavad-Gītā*, influence the thinking, attitude and inter-personal relations of human life. The relevant *śloka* is given below:-

त्रिविधा भवति श्रद्धा देहिनां सा स्वभावजा ।
सात्विकी राजसी चैव तामसी चेति तां शृणु ॥

— *Bhagavad-Gītā* (17.2)

Human nature has a variety of dimensions (Bhattacharya, 1937), and Caraka (Charak), the ancient doctor, who expounded the Āyurvedic system of medicine and analysed the human constitution and classification of human nature as a psycho-biological complex with a view to discovering the aetiology, of bodily and mental diseases and their remedies.

Caraka elaborates the several mental types bearing with the three basic elements of *triguṇa* as referred above.

According to *Bhagavad-Gītā*, it is the nature of the innate faith, types of worship, sacrifice, religiosity, austerity, penance, gift and charity a man makes, and the type of food he relishes, which determine the character and life of the people.

In this context it has been found that man's nature conforms to the kind of actions he performs. Generally, it may be either sāttvic, rājasic or tāmasic. As nature (*guṇa*s or modes of *Prakṛti*) is imbedded in the mind, his frame of mind is believed to reflect his nature (Goyanka, 1978).

The distinctive characteristics of a sāttvic man coincide with the "divine properties" and that of a tāmasic man with the "demonic properties," and the rājasic man come in between the two and may be blended with either sāttvic or tāmasic dispositions.

Foods which promote longevity, intelligence, vigour, health, happiness and cheerfulness, and which are sweet, bland, substantial and naturally agreeable, are the sāttvic type of men (*Bhagavad-Gītā*, 17: 8).

Foods which are bitter, acidic, salty, overhot, pungent, dry and burning and which cause suffering, grief and sickness, are the rājasic type of men (*Ibid*, 17: 9).

Foods which are half-cooked or half-ripe, insipid, putrid, stale and polluted, and which are impure too, are the men of tāmasic disposition (*Ibid*, 17: 10; Sahay, 1998: 138-41).

Life: Āyurvedic System

Life has direct bearing with the health of a person. The traditional system of treatment is Āyurveda which in the past had created medical marvels by renowned *vaidya*s like Caraka, Dhanvantri, Suśrut, Vāgbhata and the like. Some of them also operated on the brain of persons. Now this old system of treatment is fastly being revived. Āyurveda had conceived that a person should live for one hundred years if certain precautions are properly heeded. Here, I would like to present a few "nectar of words" in the form of *śloka*s which are given in the *Āyurvedīya Hitopadeśa* (Kulkarni & Desai, 1955). They are as follows:-

न चैव ह्यास्ति सुतरमायुर्वेदस्य पारम् ।
तस्मादप्रमत्तः शश्वदभियोगभस्मिन् गच्छेत् ॥
— चरक

(The Āyurvedic system of treatment is nothing beyond this world. That is why one should not go for a thing which goes beyond it — Caraka).

भिषजां साधुवृत्तानां भद्रमागमशालिनाम् ।
अभ्यस्तकर्मणां भद्रं भद्रं भद्राभिलाषिणाम् ॥
— वाग्भट

(The mode of habit of the *Sādhus* (*sādhaka*) are in fact, medicines. It is also a fact that preventions are better than medicines — Vāgbhaṭa).

The following *śloka* wishes the smooth function of the body of a person:

ऊर्वोरोजो जङ्घयोर्जवः पादयोः प्रतिष्ठा ।
अरिष्टानि मे सर्वाऽऽत्माऽनिभृष्टः ॥
— *अथर्ववेद* 19/61

(Be strong in arms, thigh would be firm, legs work properly, feet may be stable (there should not be paralytic), all parts of the body should function properly, and the spirit in life (soul) be vigorous — *Atharvaveda* 19/61).

The following *śloka* refers to the responsibility of the person who maintains the ideal situation connected with the functioning of his body:

सर्वमन्यत् परित्यज्य शरीरमनुपालयेत्।
तदभावे हि भावानां सर्वाभावः शरीरिणाम्॥
— च० नि० 6/7

(The clever man should take good lifestyle which could help the body so that it could prevent diseases and increase the strength by doing proper wisdom and observances — *Caraka Nighanṭuḥ* 6/7).

Dharma, Artha, Kāma and *Mokṣa* and behaviour of manliness (goal of man) — all depend on the source of body. One could spare everything except the protection of body because without it the body won't exist.

Life: Indian System of Astrology

The Indian system of astrology gives us an understanding of the human dimension of life, and also tries to give some explanations. It tells that the position and movement of various celestial bodies (planets) of the universe as they exist at the time of birth of a child, wield a life-long influence on him and go a long way in determining his temperamental disposition, inclinations and basic character. A study of such astrological characteristics in an individual among other things helps a great deal in understanding and interpreting his attributes.

Astrologically and in the traditional set-up, the temperaments of man may be divided into four *varṇa*s (literally meaning shade or colour), viz. *vipra* (brāhmaṇa), kṣatriya, vaiśya and śudra. This classification refers to the predominant individual and personal attributes of man which is considered to be universal and cuts across the four *varṇa*s of men of the same names grouped under the Hindu social system. The *vipra* represents the sāttvic taste and temperament of man, the kṣatriya the rājasic, the vaiśya a mixture of rājasic and tāmasic, while the śudra, tāmasic.

Another astrological classification of man from *Gaṇa*'s point of view refers to three categories, viz. *Devgaṇa* (*gaṇa* of

God) which is predominantly sāttvic; *Rākṣasagaṇa* (*gaṇa* of monster), predominantly tāmasic; and *Nargaṇa* (*gaṇa* of man), predominantly rājasic.

The third classification relates to *rāśi*s and *nakṣatra*s referring to distinct constellations of stars in different parts of the sky acquiring different formations. The whole sky has been divided into 12 *rāśi*s and 27 *nakṣatra*s and the persons associated with these *rāśi*s have been further classified into three major groups on the basis of the three basic temperamental characteristics of man i.e., sāttvic, rājasic and tāmasic which also speak of and determine the disposition of man (Srimali, 1972, 34-5; Sahay, 1998, 87-8).

Biological Evolution and Consciousness of Man

Evolution means the gradual development from a simple to a more complex form. Evolutionists like Darwin explained the origin of species. But in spirituality, man goes a long way in determining his cultural, social as well as biological dimensions.

The Integral Yoga of Sri Aurobindo (Donnelly, 1957:56) says that "Scientific research into the nature of subtle sheaths and of their ultrasonic rates of vibration has made progress in the West in recent years, and the following passage gives a simple and easily understandable description of their function:

"Its (the etheric body's) chief activities are two: (i) to act as a subtle matrix for all physiological changes, as the organizer wherein the pattern of the new structure lies, as a vital centre whence directive energies are variously distributed through the more mechanical chemical components of blood, organs and tissues; and (ii) to act as a bridge mechanism especially adapted to link the forces of mind, feeling and will, characteristic of each soul, to its body. A "live" body possesses such a bridge mechanism and vital matrix. In a dead body the bridge has been broken. The subtle body possesses seven main centres, called *cakra*. . . ."

Aurobindo (1970:4) says, "the evolution of Life in Matter must have been produced and governed not by a material principle, but by a Life-principle working in and upon the conditions of Matter and applying to it its own laws, impulses, necessities. This idea of a mighty life, other than the material principle, working in it and upon it has begun to dominate the advanced thought of Europe. . . ." (Sahay, 1998: 91-3).

In this context, we may refer to a case of ancient "ape-man" in which Father M. Hermens (1955: 64-7) tells that the palaeolithic cave paintings nearly thirty thousand years ago or more, were unique, marvellous depicting bisons, horses, mammoths, stags and boars. It portrayed with wonderful skill and beauty and "they reveal in the Palaeolithic man an artistry which might well awaken envy in the breast of a modern artist?"

Here it may not be out of place to mention in passing some "fascinating historical puzzles" and archaeological evidences of some ancient but advanced civilizations which properly existed in the Americas thousands of years before Columbus, and elsewhere, and the legendary lost Atlantis, a large island empire in the Atlantic to the west of Europe, which also finds mention in the writings of Plato in the fifth century BC (Berlitz, 1972, Sahay, 1998: 230).

Purpose of Life

The purpose of life depends on the different economic, social and religious backgrounds, family upbringing, surroundings, relevant events or incidents in life, source of inspirations, and the like. There are people who if placed in a given situation, may become complacent. Others who have cherished a desire and want to go up but are lazy, or want to do good things and have imagination, but resources or time do not permit. There are people who are capable and resourceful, but still they continue to be strugglers, become unsuccessful and get frustrated. There are some people who are imaginative but not practical. Again, there may be honest or dishonest, a simple man, bad man, exploitor or hypocrite, etc.

There are meritorious who get successful. But it does not mean that others who were not successful, were not meritorous. It is a lopsided thinking to say that a man who won, won only because of merit or labour. A man, who in spite of merit and labour fails, luck or bad luck it constitutes the supreme factor which must not be ignored.

There are certain things in life which are explicit, but there are others which are enigmatic. There are many people who think that surroundings are important. But others talk about heredity. But still they are not conclusive. Among the people who have a religious background with proper socialization, one of its members may become an atheist, while on the other hand a man who is from an inferior or corrupt background may gain saintly attributes. We have no logic or explanation. To understand it, we may look into the "past life," "theory of *Karma*" or the 'Unseen Supreme Factor' (USF) (Sahay, 1998: viii).

There are people who are thinkers, sober and take their life seriously and they want to contribute to the society, but they do not expect anything from the society.

There are people who have wordly problems and they go to temples or pray to God to help them. Still there are some people who worship religion or do spiritual practice, but they have no personal self. For them "means" is an "end" in itself. Here we specifically may refer to the higher level of *Yoga* or *Brahmavidyā* (spiritual refinement) in which *Ātmā* (soul) gets united with the *Paramātmā.*

Horizontal Man vrs. Vertical Man

When we think of man and his life, it may be interpreted as "horizontal man" and "vertical man." Those who are involved with mundane things, have marginal, or negative attributes, and whose activities and motivations are diffused reflect the horizontal man. But when the person is spiritual, he has spiritual attributes, or there are certain things which determine the spiritual "link-up" (spiritual communication) with God and I call it as the "Unexplained Supreme Factor" (USF), he is known as the vertical man (Sahay, 1998: 204-9).

In this context, we may have a beautiful *śloka* of *Kaṭhopaniṣad* (Nikhilananda, 1948: 33). It is as follows:

श्रेयश्च प्रेयश्च मनुष्य मेतस्तौ सम्परीत्य विविनक्ति धीरः ।
श्रेयो हि धीरोऽभि प्रेयसो वृणीते प्रेयो मन्दो योगक्षेमाद्वृणीते ॥

— *कठोपनिषद्*, 1/2/2

The *śloka* tells that everybody comes across *śreya* or *preya* paths in life. Those who are intelligent and knowledgeable by analysing the pros and cons of these two paths, will prefer to choose the first one, the excellent path of true welfare, *śreya*, than the second path of wordly pleasures, *preya*. But low intelligent people choose and adopt only the second path of *preya* from the view-point of fulfilling their wordly desires and lust. It implies the vertical man and the horizontal man respectively.

In between the *śreya* and *preya* paths in life, there is a range of ten basic gradations of the manifestations of USF which have been arranged in vertical stages. The first one being the *highest*, i.e., the *Para Brahma* (unmanifest) or the USF, while the *lowest* stage being the *Lower spirit, Ghost, Demon or usually malevolent*. In between the other stages are creator God or Brahmā (manifest), Tutelary God, Power-cult, Forces of Nature, Ancestral spirit, etc. Each of the corresponding facets is associated with spiritual or religious practices, explicit goals, experiences or *anubhūtis* (Sahay, 1998: 99-100).

Śreya path in life may gradually lead to *Brahmavidyā, Ilm Ruhānī, Vihaṅgam Yoga, Sahaja Yoga* and the like if one gets contact with a *Sadguru*. A *śloka* of *Kaṭhopaniṣad* (1/2/7) reads:

श्रवणायापि बहुभिर्यो न लभ्यः शृण्वन्तोऽपि बहवो यं न विद्युः।
आश्चर्यो वक्ता कुशलोऽस्य लब्धाश्चर्यो ज्ञाता कुशलानुशिष्टः॥

It means that there are many who do not even hear of *Brahmavidyā*, though hearing of it, many do not comprehend. Wonderful is the expounder, the rare the hearer; rare indeed is the experiencer of this knowledge taught by an able preceptor (Nikhilananda, 1994: 133).

When a vertical man gets elevated, it may lead to mysticism and there is a large body of different religious literature having a bearing on it (Sahay, 1999: 61-148). Some of the terms are symbolic. Kabir Das tells:

नैया में नदिया डूबी जाय।
चूँटी चली अपने नैहर को नौ मन काजल लाय।
ऊँट को बगल तर लीन्हा हाथी लियो लटकाय॥

गुरू ने पैठाया चेला, नयामत लाना
पहली भिक्षा अन्न का लाना
गाँव शहर के पास न जाना
हिन्दू मुस्लिम छोड़ के चेला
झोली भर के लाना.....

Mahayogi Gorakhnath (Ibid, 1999: 126) tells:

गगन मंडल में औंधा कूवाँ, जहाँ अमृत का बासा।
सुगरा होई सो भर भर पीया, निगुरा जाय पियासा॥

Goswami Tulsi Das Tells:

बिनु पग चलै सुनै बिनु काना, कर बिनु करम करै विधि नाना।
आनन रहित सकल रस भोगी, बिनु बानी वक्ता बड़ योगी॥

Paramhans Dr. Chaturbhuj Sahay tells:

कल तलक था मैकदे में खुश्क साहिल की तरह।
आज साकी ने मुझे कतरे से दरिया कर दिया॥

Christianity also seems to have developed the spiritual techniques of link-up with the highest Spiritual Being as reflected in several passages of the Bible (Sahay, 1998: 88-89): Two of them are given below:

> The Light of the body is the eye: if therefore thine eye be single, the whole body shall be full of light.
>
> (Mathew/6-22).

> And I heard a voice from heaven, as the voice of many waters, and as the voice of great thunder: and I heard the voice of harpers harping with their harps.
>
> (Revelation 14.2).

When a man becomes elevated there is the awakening or activation of the *Kuṇḍalinī* (serpent power) and *Suṣumna*, and some supernatural or miraculous things may come into being.

Entry Point of Spiritualism

Mundane and religious or spiritual life always may not be a water tight compartment, but mundane things with experience or age ultimately lead to spirituality as it is also known in the traditional system of the four *Āśrama*s, viz. the *Brahmacarya* (studying), *Gṛhastha* (householder), *Vānaprastha* (retired life), and *Saṁnyāsa* (complete renunciation) (Prabhu, 1940: 84).

In this context, generally the last two *Āśrama*s relate to entry point of Spiritualism and its Awareness. Here there are different relevant points, but I have delineated seven points which I call to be the most basic. They are as follows:

(1) *Theism* (functional): It is "functional" because it is not just the faith, but it is effective and it really works; (2) *Purity*: There are two types of purity: physical, and thinking. At the moment, it refers to cleanliness and bathing, abstenance of alcohol, taboo food, free from pollution and defilement. If it is so, in due course it may switch over to the 'purity of

thinking'; (3) *Temperance*: It refers to moderate alcoholic liquors, avoiding extremes of either sex or other sense of control over sense-organs; (4) *Love to all Beings*; (5) *Charity*; (6) *Honesty*; and (7) *Discipline*. (Sahay, 1998: 171-72).

Spiritual Postulates Bearing with Life

Whatever the higher society works, it has certain norms and pre-conditions. I have tried to refer to 18 basic selected spiritual postulates for the guidance to the human life. They are: (1) Man to man (Social), (2) Man to himself (Introspectional), (3) Man to USF (Link-up). The further details are given below:

MAN TO MAN

(i) Be humanitarian. Help the needy people if you are in a position to do it.

(ii) Privilege and duty are interdependent.

(iii) Parents must give positive socialization to the children.

(iv) The parents should transmit proper *Saṁskāra* to the children.

(v) A human being has "merits" as well as "demerits." But one should concentrate on the first one, while the second one should be ignored.

(vi) Science believes that *action* is followed by "reaction." It works also on a higher and spiritual level. If you do good/bad thing — it returns to you.

MAN TO HIMSELF

(vii) Develop your awareness of consciousness in such a manner that you could enlarge not only the neighbourhood, area, state, region, nation, but also the whole world.

(viii) Positive thinking is good, but it must be followed by *action*.

(ix) Develop a philosophy of life: be systematic and methodical, make advance planning, allow several alternatives to avoid frustrations.

(x) Psychological counselling is necessary to locate talents and creativity of the people.

(xi) Make your own self-estimates and critical evaluations, and modify yourself.

(xii) People generally identify to a caste, sub-region, region, or a political and religious group, but hardly they properly identify and concentrate on the name of the person itself which has "positive vibrations" or energy which if one properly taps, it will give a new dimension to the personality. Such names are connected with gods, divinities, saints, heroes, mythological characters, positive attributes, viz. Ram, Krishna, Mahadeo, Surya, Ajit, Kalyan, Dharmnath, Anil, Kiran, Atal, Jaswant, Sajjad, Hamid, Ajhar, Altaf, and the like.

MAN TO USF

(xiii) Evoke the USF (Unexplained Supreme Factor) and pray for a personal thing. But you should not insist or dictate. The USF should be given "options."

(xiv) On a higher level, USF never does any wrong. It is man who errs. You must get convinced gradually.

(xv) Be a good "actor" in a "drama of life" and you should thoroughly learn by heart the "script" given to you. Perform your proper duty, but be detached because things are not within your control. It has bearing with a *śloka* of the *Bhagavad-Gītā* (II/47).

कर्मण्येवाधिकारस्ते मा फलेषु कदाचन ।
मा कर्मफलहेतुर्भूर्मा ते सङ्गोऽस्त्वकर्मणि ॥

(Your right is to work only, but never to the fruit thereof. Be not instrumental in making your actions bear fruit, nor let your attachment be to the action).

(xvi) Pray to USF every morning (or when he is calm) so that USF could help "everybody," "the whole world" irrespective of good or bad. Thus, you can create (positive vibrations) which is useful in the society.

(xvii) The spiritualist should work like a *panihārin* (woman carrying pitcher over her head full of water) who goes around, talks with people (mundane things), but still she is highly cautious and balanced. Any distraction of the "*surati*" (conscious power of the soul) from the "pitcher" will make it fall off the head.

(xviii) In case you think that there is discrepancy, controversy, or you have to do certain things which are unpalatable or even immoral, "tell it" mentally to the USF and be "witness" that you had to do it under unfavourable circumstances. "Confession" is necessary (Sahay, 1998: 171-80).

Models in the Muddle

A society which has a small range of social relations (Radcliffe-Brown, 1964:178-204) with simple norms, personal interactions of models, is homogenous and has a unity of thinking. But when it leads to a long range of social relations, it becomes complex, heterogenous and such models become diverse. In the same set of models they may have alternatives, but the two may have some different implications. Again, some models cut across, may be controversial or relative. A model has bearing with early socialization, upbringing of a person, surroundings, profession, religions, political affiliations, source of inspiration and the like. At this juncture of time, there is tremendous development of communication, global multiplicity, and a large body of literature connected with social models, but it leads to a sort of "model explosion." It creates confusion, conflict and disorganization. Life is fast changing, there are marked generational gaps, values are decaying, and traditional and modern models do not work together. They constitute a serious problem of tension and adjustment.

Jesus Christ said that a person who was slapped, should present his other cheek for another slapping. Mahatma Gandhi drove out the Britishers from India through *Satyāgraha* and non-violence. Other revolutionaries of the time had the model of violence for the same purpose. There are some people who do service to *Sādhus* who may be "fake." But still, there may be a different model in which once when "*Mahāvtar* Babaji," the *Guru* of Sri Lahiri Mahasaya, during the *Kumbh* at *Triveṇī Saṅgam* was serving a saint, Lahiri ji wondered as to why that Babaji should serve this saint who was just an ordinary person? Babaji said that the saint might be ordinary or elevated, it did not matter, but the service which he was rendering to that saint, the "devotion of service" through that saint, was being received by God Himself. Arjun when he was on the battle field of Kurukṣetra and laid down his Gāṅdīva, was his own model of thinking. But Lord Krishna presented him another model of *Karma Yogī*. There is a model in which respected saints travel taking proper tickets. But there was another model in which a highly respected saint of south India with good following, formerly used to travel on train without a ticket as he thought that he did not have money to buy a ticket, he was a recluse and just doing a "divine service to the people." There are a number of other *sādhus* or saints, of course who are genuine, follow this model though it is dangerous. There is a model in which a person who is sensitive cannot harm a creature, but there is another model in which there are *thugs* who harm, or kill a person easily by snatching his belongings. Similarly, there are a number of "good" and "bad models" and many in between the two. In olden days, people were mostly involved with positive attributes, had commitments, had a philosophy of life, but now many people have become shaky, are in doldrums and their conscience has dimmed. Such people have become self centered, think of manipulations creating a new type of model and perpetuating it with their own justifications.

Life and Modern Science

Modern science has given a number of boons to the society and they have enriched human life. Now telephones and mobile phones are so easy that you can contact a person from the longest distance in the world within seconds through satellite. It can be used in a car, train or plane which earlier was not possible. So is the case with computer which has become very popular. By and large, and even children are using different types of educational and recreational programmes and varied global information through Internet, chattings and even bookings for train and plane reservations. Similarly, now televisions have nearly two hundred channels covering different countries. It is heartening that telephone, mobile and television have gone into many rural areas.

In recent years night-soil has been explored, and a company has developed cheaper latrines on mass scale and is generating power for street lights. This unique institution is known as "Sulabh International" which works not only in different parts of the country, but some neighbouring countries as well. Besides, cow dung is producing *gobar gas* which is used for domestic lighting, *culhā gas* and even incense (*agarbattī*).

Solar energy is still another aspect which is being used for various purposes and it has become very popular throughout the country — urban as well as rural areas. It does not only generate electricity for lights, but is used also in different ways, for heating of water, for air conditioning, solar cooking and the like.

However, in spite of scientific developments which have improved the quality of life, there are certain aspects which are highly distressing. One of the aspects is the disease of AIDS which is spreading not only on a country level, but throughout the world, and people are trying to control it through preventive measures on a war footing. Besides, population explosion which has a number of side effects like unemployment, *bandha*, indiscipline, crimes, terrorism, accidents, diseases, cutting of trees, pollution, warming of atmosphere, ozone problem, radiations, lowering of ground water, and the like, is directly and indirectly affecting the quality of human life.

References

Allen, R.E. (1990). *The Concise Oxford Dictionary*, The New Edition for the 1990's (Eighth Edition), Bombay (Mumbai): Oxford University Press.

Aurobindo, Sri (1970). *Evolution*, Pondicherry: Sri Aurobindo Ashram.

Berlitz, Charles (1972). *Mystries from Forgotten Worlds*, London: Souvenir Press.

Bhattacharya, Haridas (1937). *The Cultural Heritage of India*, Volume III, *The Philosophies*, Calcutta (Kolkata): The Ramakrishna Mission, Institute of Culture.

Chaturvedi, T.N. (1989). Editorial, *The Indian Journal of Public Administration*, (Special Issue), vol. 35 no. 3 July-Sep., p.ix.

Donnelly, Morwenna (1957). *Founding the Life Divine:An Introduction to Integral Yoga of Sri Aurobindo*, Bombay (Mumbai): Jaico Publication House.

Goyanka, Jayadayal (1978). *Śrīmad Bhagvad-Gītā*, Gorakhpur: Gita Press.

Hermanns, Fr. M. (1955). *The Evolution of Man*, Society of St. Paul, Allahabad.

Howard, Henery (1952). *Happy Life.*

Inamdar, N.R. (1989). Environmentalism and Development, The *IIPA* (Special Number), vol. 35, no. 3, July-September, pp. 353-61.

Imam, Bulu, (1990). Ecological Background of Chotanagpur : Its Imbalance and Tribal Communities, *Man In India*, vol. 70, no. 3, September.

Kulkarni D.A. & Desai R. (1955). *Ayurvediya Hitopadesh* (Hindi), Calcutta (Kolkata): Baidyanath Ayurveda Bhavan.

Nikhilananda, Swami (1949). *The Upaniṣad*, vol. I, New York: Bonaza Book.

Prabhu, H.N. (1940). *Hindu Social Organization*, Bombay (Mumbai): Popular Prakashan.

Prasad, R.C., *Kalaapi*, N.D., (Hindi).

Radcliffe-Brown, A.R. (1964). *Structure and Function in Primitive Society*, London: The English Language Book Society & Cohen & West Ltd.

Sahay, K.N. (1998). *Spiritual Anthropology : The Spiritual Dimension of Man*, New Delhi: Commonwealth.

Sahay, K.N. (1999). *Social Anthropology in India*, New Delhi: Commonwealth.

Sahay, K.N. (1999). *Beyond the Thousand Petalled Lotus and the Science of Consciousness*, New Delhi: Commonwealth.

Shakespeare, William (1952). *We Are Such Stuff.*

Shakespeare, William (1952). *All The World's Stage.*

Srimali, Narayan Dutt (1972). *Bharatiya Jyotish* (Hindi), Delhi: Anupam Pocket Book.

Verma, Mahadevi N.D., *Virah Ka Jaljat Jiwan* (Hindi).

5

Life Perennial

Jogananda Jha

ĪŚHOPANIṢAD says whatsoever exists in the world is the abode of Iśa — the ultimate power, the cosmic energy. So, life cannot be anything except a part of that cosmic energy. This life energy is perennial, it is imperishable. *Kaṭhopaniṣad* says — the seer is not born nor does he die.

The wave of life, like any other energy, is continuously reacting with other forces in the environment, while trying to settle at a suitable space and time. Finding a suitable space it expands at the cost of tiny particles surrounding it, imbibing them, absorbing and assimilating them like the unicellular amoeba, whose vegetative activity of engulfing the surrounding particles and expansion can be observed under the microscope. Thus it takes a material form; may be the form of a fertilised ovum in the womb of the mother in the form of a condensed mass of energy, with all the genetic codes from the ancestors embedded on it. This brings us to the subject of Molecular Biology or Genetic engineering where the Kinetic energy can be observed and its direction can be controlled. This mass of energy is incessantly growing at the cost of the surrounding materials, taking the form of an embryo and the fetus, till the womb is no longer spacious enough to accommodate and retain it. So, the life is born in the shape of an infant.

And since the infant is nothing but a condensed mass of energy, its different parameters, its potential expansion in space and time can be calculated and speculated. That is how a good astrologer can predict the whole future life, taking in view its interaction with the different waves of energy emanating from the sun, the moon and other constellation of stars at the exact moment of birth.

Now this energy in the physical form of a child is incessantly amalgamating, absorbing, and assimilating matters from the atmosphere and goes on growing till its full maturity at youth. In this course it utilises the useful matters for its growth and rejects others by excretion of faeces, urine and sweat. When this energy is fully saturated, it is no more in a position of absorbing anything from the external atmosphere, the life maintains a status quo for a certain period of time. Nevertheless a struggle between the centripetal and centrifugal forces continuously goes on and ultimately the life energy begins to yield to the external forces, gradually losing all its external attainments in the old age till it is no longer able to hold and retain its physical body. So the body perishes and the life is liberated. The man with the name and form of the body is dead but the life or the soul is alive.

Now, as stated earlier, this liberated life wave in the celestial form is in search of a suitable space to stabilize and unfold. If this energy wave has many twists and turns due to its interactions in its earlier stage or past life, it prevails over different celestial layers called *Loka*s or *Yoni*s, the *Pretaloka*, the *Pitr̥loka* from where we can get a glimpse of image off and on. This very wandering stage of the life energy can give rise to rebirth and expands the theory of rebirth. Jesus Christ got up on the third day and the day of resurrection is believed by both Christians and Muslims alike.

In case the energy wave is not twisted, nor complicated, it simply merges harmoniously into the all-pervading cosmic energy like the last fading musical sound merging into the atmospheric sound — there is no rebirth, the soul is liberated, it gets *nirvaṇa*, it gets *mokṣa* or Salvation.

So, *kurvānweh karmani, jijiwiśet satyam sama evam twayi nanyathelo-asti, na karm lipyate nara.*

6

Web of Life

Ajeya Jha

This we know: the earth does not belong to man, man belongs to the earth. All things are connected like blood that unites us all. Man did not weave the Web of Life, he is merely a strand in it. Whatever he does to the web, he does to himself.

— Chief Seattle

WE have deliberated upon life and lives. But there is another vital issue namely the exact nature of relationship existing between different life forms. Are they related to each other in any way? Is the sum total of all the lives in this world just that much? If we remove one life from an eco-system, will the loss be of just one life? Or more? Or less? Is the ecological equilibrium that we talk about real or does it have its own dynamics?

I am deliberating on this aspect not because of the recent obsession with biodiversity. I rather have been inspired by our own ancient Indian wisdom — *Live and Let Live*; by the age-old Buddhist tradition of *Ahiṁsā* — Non-violence. We need to ponder important questions over here such as why did Mahatma Buddha consider other lives — even those so-called most insignificant ones also, as sacred? Why should the Hindu Mind consider living at the cost of other lives as morally wrong and sinful? What is the logic to let others live? How does it help us if we promote other lives too? Was it prophesized only on moral/emotional grounds or does it have a much deeper implication for us?

Web of life is the sum total of all lives — all kinds of lives — the symbiotic relationship that binds them, plus their vibrant interaction with the earth, the Sun, the rains and the air. This web of life is often understood in terms of the wide variety of plants, animals and micro-organisms. So far, about 1.75 million species have been identified, mostly small creatures such as insects. Scientists reckon that there are actually about 13 million species, though estimates range from 3 to 100 million.

In 1996, the World Conservation Union (IUCN), in collaboration with over 600 scientists, assessed the status of animal life on earth. According to their estimates, 25 per cent of mammal and amphibian species, 11 per cent of birds, 20 per cent of reptiles, and 34 per cent of fish species surveyed so far are threatened with extinction. In addition, another 5 to 14 per cent of species in these groups are nearing threatened status. Further, scientists estimate that today *extinction rates are many hundred times greater than normal, and are mounting up in a geometrical progression*. At the Rio Plus 5 summit at the UN in 1997, a report estimated that approximately 50,000 species of flora and fauna were disappearing annually. It must be even higher now. If the trends keep on to intensify, as the evidence suggests, it is estimated that almost 70% of flora and fauna will be dead by the end of the next century. One may argue that when with so much of loss we have not been affected then why should we bother with the loss of Life? Life does not follow barometric principle. A loss of 10% does not mean a 10% ruination of human life. Rather it may have no consequence at all, yet the next 0.01% damage may spell our doom. There is a threshold of biodiversity below which life or even the possibility of life is reduced to zero.

Is this a new phenomenon? Is this massive loss of life being witnessed for the first time? Scientists have suggested that earlier in the history of earth extraordinary loss of life had occurred as many as five times. Life on earth evolved for approximately four billion years from its origin to the present. Five times during this period mass extinction of life occurred.

1. Nearly 85% of species were lost about 438 million years ago during the Ordovician-Silurian Extinction.
2. Again almost 85% of species were lost about 360 million years ago during the Devonian Extinction.
3. The Permian, 250 million years ago, saw the loss of more than 90% of species — the worst ever.
4. 76% of species were lost during the Triassic period, 210 million years ago.
5. The last of these five catastrophe which wiped out the mighty Dinosaurs occurred 65 million years ago during the Cretaceous period.

An interesting point to note is that all these occurred well before the arrival of *Homo sapiens* or even the primates in general. In other words this loss holds no assurance for us. We are rooted in the web of life as of today and the loss or gain of life that occurred earlier is irrelevant for our survival.

Why did such massive loss of life occur? Scientists are not sure but the natural disaster such as meteorite impact and climactic transformation are believed to be responsibe for this. Thus of about the one billion life forms that had existed over our earth, 99% have disappeared.

Is it a very assuring thought? Does it absolve the humanity for the huge loss of life occurring now? If life has been destroyed by nature itself why should we feel guilty about the loss we are causing? A question often asked is: why do we need to worry about other species? What is wrong with having the earth which is directed towards just the well-being of human-beings? Why should not we change our earth so that human prosperity is eternal? Aren't we being unnecessarily emotional by worrying about other animal life?

The clear distinction between the losses of life that occurred earlier and that which is taking place now needs to be understood. To begin with, this is the first time a creature (we the human beings) of the earth is responsible for the losses. This matters because nature, we know, has a method in its madness, whereas destruction we are causing is more of madness in madness. Secondly, during previous disasters it was just the existing life that was affected. Now future life too is being affected. During the Cretaceous extinction which saw the disappearance of Dinosaurs, an obscure group of mammals survived. In time to come it revived the tradition of life on earth and today mammals are the major forms of life. If that obscure mammal too had died, we, the human would not have been there at all as we too belong to the family of mammals. Today the destruction is anthropocentric — directed towards the well being of humans — just one species. Whatever appears useful to us survives; the rest has to perish. But is that what appears useless to us really useless? Most species are unknown and unmonitored. 95% of biodiversity we are losing is a mystery to us. We know nothing about it. *The weed of today could be the need of tomorrow.* We are destroying many plants which hold the key to cure future diseases. If we eliminate biodiversity we may be eliminating the chance to discover life-saving drugs. We are exterminating sources that could be the only possible food for us tomorrow. We are inflicting changes that will rob us of vital oxygen or bring about temperature changes that will wipe us off. What is more, we are murdering the earth's ability to revive life in future. It is this slaying of future life that prompted William Butler Yeats to conclude — *Man has created death.*

Man's greed has led to a situation where the trauma of extinction of various forms of life is being played out on a fast-forward mode. The pace of transformation is mind-boggling. Scientists fear that the complete web of life is caught in a countdown. Those who understand the process of evolution believe that no life is escaping the dragnet of the human greed. The earth without life is not just inevitable but imminent too, believe the scientists. Are we — the humans — a tool to bring life to oblivion? If so, will nature give life a chance ever again?

Why has this happened? Why happened now when our ability and achievements are at their zenith? What change has taken place in human consciousness? The following may appear to have brought us to this predicament:

- Estrangement with our traditional wisdom, values, and moral principles that universally held nature as sacred, divine and supreme, and accorded prominence to its inviolability.
- The belief that humans are superior to other life-forms and the dogma that nature exists merely to serve human will and purpose. The conviction that 'nature is savage' needs to be changed.
- The dominant economic and development prototypes of the modern society, which lay highest importance on the values espoused by market, and perhaps least on nature — where gold holds a higher standing over life.
- Industrialization, economic globalization, cultural homogenization, commodity accumulation, and urbanization. Each of these ignore that earth is finite and thus contradict the continuation of Life.

My candle burns at both the ends
and it will not last the night;
But Ah! my friends and Oh! my foes
it gives a lovely light

- An unflinching reverence for technology and a limitless faith in the rightness of science. The belief that technological evolution is ethical, ideal and inevitable. Our tendency to lash out anyone questioning such a conviction. Our purposeful blindness to the technology that is reduced to being the weapon in the hands of the oppressor.

What were the earlier values and ethics of behaviour that we have abandoned? The practice of conscientious living — in harmony internally and externally — was the essence of philosophy all over the world. Tribal beliefs, Vedic tenets, Islamic and Christian doctrine and above all the Buddhist and Jain philosophy all teach us to respect life.

> As I am, so are these; as these are, so I am; identifying himself with others, let him not kill nor cause any one to kill.
>
> (*Dhammapada*)

> Nature when violated, truly destroys. Nature when preserved, preserves. Nature, therefore, should not be violated, lest the violated Nature destroys us.
>
> (*Manu*)

> All living beings whatever, without exception, weak or strong, whether long or high, middling or small, subtle or gross, seen or unseen, dwelling afar or near, born or yet unborn: may all beings gain inner joy.
>
> (Buddhism's *Suta Nipata*)

> Refrain from taking precious life from any living being, even to save your own life.
>
> (*Tirukkural*, Ancient Tamil scripture)

Do we need to respect other life? Do we need to care for it? We must, if we do not wish our existence cut short. It is the stirring symbiosis of life forms and their interactions with each other and with the rest of the environment that has made earth a uniquely habitable place for humans.

But our values that revere nature had a higher philosophical purpose. Ancient minds were not upset with the possibility of human extinction. To them continuation of life in any form was the higher ideal. They were not disturbed by the truth that earth is not for humans alone. It is for all life — especially the future life. That continuation of life is more important than the survival of an individual or even an entire species. In India we were familiar with the eternal cycle of *Satyuga* and *Kaliyuga* — of evolution and degeneration, yet we did not dilute our conviction of reverence for life — our supreme obligation to support the continuation of evolution.

Because the fate of every species is to become extinct eventually, there must be an influx of new species for life to continue. This is possible only when the present web of life is allowed to exist and not devastated by myopic human arrogance. This is because biodiversity adds to the resilience of life. If and when changes will occur — a new disease is introduced — many will die but some will survive. Even recently Jerry Brown, a noted philosopher, commented:

> We must always remember that we are but a small point on a long line of time God has allowed us to extend. Small as it may be, it is ours . . . ours to fill with purpose and meaning, ours to make honourable and distinguished . . . and ours to defend in the court of history. (Jerry Brown)

The root of the current crisis lies in our misguided enterprise, our fallacious endeavour. Beyond biological evolution was the intellectual evolution which logically should have gone hand in hand with spiritual evolution and not the materialistic evolution. From Buddha to Mahavir to Christ to Mohammed to Gandhi to Teresa — all sincerely preached spiritual quest alone. We revere them, yet we ignore them. We admire their ideals, yet do not consider them worth emulating.

Life is sacred. Life is the only ideal that is to be celebrated. Once we surrender to this fundamental principle, we can regain our privilege to support eternity of life — of evolution. We shall rejoice in simple living and will honour minimal consumption. We will give due importance to values other than the market values and will acknowledge that nature and its richness are the ultimate measure of our intent and success. We will look upon nature as a guardian and not as a commodity.

Yes, life does have a future. Future as a web of life and not as a stand-alone life.

7

Life: Christian Perspective

Anthony Sudil

Introduction

THE term Life denotes the organic functioning of plants, animals and humans. Life and death are the opposite categories basic to all living things; both categories relate to and interpret the other. Human life is unique. It is not merely instinctive but is capable of self-realization and open to formative influences. There is a radical distinction between life as a vital, natural force, and human life which has a stronger ethical content and means and manner of life with a definite goal. The Bible, which is the scripture of the Christians, tells us that human life is a fellowship with God which humankind enjoy as a specific gift from God himself.

The Christian perspective of life is based on Bible and Biblical traditions. The Bible tells us that life is sacred and is created and gifted to humans and other living beings. As God is the author of life, no one has the authority to misuse life according to one's whims and fancies. Life has to be lived and used according to the specific plan and purpose of God which he has put into the created world.

Literal Meaning of the Word Life

The English noun 'Life' is derived from the Latin Verb *vivere*, meaning "to live"; it denotes life in its concrete outward manifestations. It is used generally for a lifetime or duration of life and specifically for an individual's way of life. It is not a once-for-all historical event but as a type of super temporal behaviour existing alongside other possible types.

Meaning of Life in Greek Philosophy and Literature

Christian understanding of life in later stages and in theological and philosophical interpretations is very much influenced by Greek philosophy and literature, especially of Aristotle and Plato. In Greek philosophy, life refers in the first place to that living quality of nature which is shared alike by men, animals and plants. For the Greeks, therefore, life belongs to the category of natural science, being characterized by the power of self-movement as distinct from mechanical movement.[1] The cause of life is considered to be the soul, which Diogenes of Apollonia imagined as an ethereal substance while Xenophanes thought of it as a fluttering breath.[2] Natural life is made up of the components soul and body. Even the Gods are imagined largely as living creatures having two natures, analogous to the human body and soul.[3] Plato distinguishes between living creatures which are mortals like the humans, and immortals like Gods.[4]

The Greek philosophy added a third element to life which is specific to only human life. This they called human reason, mind, or understanding. Human body is part of a purely natural life, but human soul and reason (mind) are divine elements which are given only to human life. This makes them radically different from animals and other animate beings. While the humans share body aspect with the animals, soul and mind

1. Plato, *Leg*. 10, 895c ff; *Phaedr*. 245c ff; Aristotle, An. 2, 2 p 412b, 16f; p 413a, 22ff.
2. H. Diels, *Die Fragmente der Vorsokratiker*, I, 425, and 42.
3. Aristotle, *Metaph*. 11, 7 p 1072b, 28 ff.
4. *Tim*. 38c ff.

give them power of various alternative lifestyles and manners.[5] The Greeks' ethical view of life is expressed with great clarity by Aristotle, who distinguishes between the following ways of life: (1) Pleasure-loving, (2) Active, (3) Political, (4) Contemplative.

The neo-Platonism differentiated between life in this world and life beyond this world. According to Plotinus, human indeed possesses natural life, but the life which is perfect and true is to be found only in the one divine realm of the One. The way to this true life leads via abnegation of the body, and cleansing from all earthly things till humans attain the true life and become one with the divine.[6] In this way they perceive life as an upward ascent.

The Early Christian Perspective of Life based on Scriptural Traditions

The Christian perspective of life in general and especially of human life is based on Christian Biblical theology of life developed in the milieu of Jewish theology and Greek philosophy. Human life is a complex reality besides its day-to-day existence in which humans' life consists in eating, drinking, sleeping, etc. For a Christian, life consists in living according to the values of Jesus Christ and his Kingdom. Ordinary food is not sufficient for such a life. It also concerns eating and drinking of spiritual food, which is belief in Christ. In John's Gospel Jesus tells his disciples that they shall not work for food that spoils, but for food that endures to eternal life.[7] It is faith in Jesus that gives real life to the humans in this world.[8] According to the belief of the early Christians, which

---

5. Demosthenes, *Orationes* 7:17; *Dionysius of Halicarnassus*, 3, 17, 7; Plato, *Rep*. 521a; *Leg*. 7. 806e; 3, 944c.

6. cf. Plotinus, *Enneads* 1, 4, 3; 6, 7, 31.

7. Gospel of John, a Book of Bible, (Hereafter, Jn) 6:27.

8. Jn. 6:33.

is recorded by John, Jesus is the real bread of life. Then Jesus declared, "I am the bread of life. He who comes to me will never go hungry, and he who believes in me will never be thirsty."[9] Real life for a Christian consists in eternal life or life in God's Kingdom which Jesus came to establish.[10] In John's Gospel Jesus assures the humankind that everyone who believes in him shall have eternal life and they shall live with him at the culminative arrival of God's Kingdom.[11] According to the Christian perspective of life, the material things we hoard up, the food that we eat or any such thing will not truly give us true life. Only faith in Jesus and abiding love for him will give us true life.[12]

Life in the Community

This abiding love for Jesus also helps the Christians to live their lives like Jesus lived his life here on earth. His life was one of sharing and so the Christians are invited to share their wealth and fortune with others.[13] Real love shows itself in the concrete act of sharing this world's goods with a brother in need.[14] The manner in which a Christian lives his/her life has to be radically different from that of an unbeliever. They have to avoid ostentatious living, which is a life of pride.[15] The believers are to be free from the cares, riches and pleasures of life. This perspective of life among Christians is in contrast with the Greek understanding of life. The reason why this

---

9. Jn. 6:35, Ref. 6.48; 53-56.
10. Gospel of Mark, a Book of Bible, (Hereafter, Mk.) 1:14.
11. Jn. 6:40 and 47.
12. Jn. 6:49-51; S.H. Hooke, Life after Death V.G.E. Ladd, A *Theology of the New Testament*, 1975, 254-69, W.B.J. Martin, Life after Death III; *The After-Life* in *the New Testament*, *ExpT* 76, 1964-65, 332-37.
13. Mk. 12:44 Lk. 21:4.
14. I[st] Letter of St. John, a Book of Bible (Hereafter 1 Jn.), 3:17.
15. 1 Jn. 2:16.

specifically Greek conception of life is not adopted by the Christian community is that according to the latter, a human does not live his/her life for himself/herself or to develop his/her own virtues, but is answerable to God for living in the service of others.[16] Life and all qualities of life which a human person receives from God are to be used for the service of other humans and life of the community.[17] Life of a Christian has to be the image of God himself. In the Christian tradition God is seen as a God of mercy and compassion; similarly the life of a Christian has to be of mercy and compassion.[18]

Life is a Gift of Creator God

In continuity with the Jewish view of life, the early Christians consider natural life as a priceless possession.[19] God is the undisputed creator of life in the universe and human life.[20] He is the Lord[21] and the embodiment of life. He is the living God[22] and the God of the living.[23] Jesus, through his ministry, gives the message that he has come to protect, preserve and

---

16. R.W. Thomas, "The Meaning of the Terms Life and Death" in *The Fourth Gospel and in Paul*, *SJT* 21, 1968, 199-212.
17. Letter of St. Paul to Romans, a Book of Bible (Hereafter, Rom), Ch. 12; 1st Letter of St. Paul to Corinthians, a Book of Bible (Hereafter, 1 Cor.) Ch. 12.
18. 2nd Letter of St. Paul to Corinthians, a Book of Bible, (Hereafter, 2 Cor.), 1:3; Rom 12:1; Letter of St. Paul to Colossians, a Book of Bible, (Hereafter, Col.), 3:12; Gospel of St. Luke, a Book of Bible, (Hereafter, Lk.). 18:1, Jn. 4:24, Rom. 8:26.
19. Gospel of St. Mark, a Book of Bible, (Hereafter, Mk.), 8:37.
20. Acts of the Apostles, a Book of Bible, (Hereafter, Acts), 17:25; Gospel of Matthew, a Book of Bible, (Hereafter, Matt.), 10:28; Rom. 4:17.
21. Lk. 12:20; Acts 10:42, Jas. 4:15.
22. Matt. 16:16; 26:63.
23. Matt. 22:32; Mk. 12:27; Lk. 20:38.

even to restore life.[24] The Gospels proclaim that true life depends on the word of God.[25] As life is a gift of God, he has given us the basic necessities, such as food and clothing to preserve and protect life. They are to be gratefully received as gifts of the Creator and used in a responsible way.[26]

Present Life as a Pointer to Eternal Life

Eternal life is a sort of participation in the divine life, since God and Christ are alone fully possessed of life and are thereby the sole bestowers of life.[27] Although the present life is precious and important there stands the life to come.[28] A good and godly life brings blessings to the present life and leads to a blessed eternal life to come.[29] Eternal blissful life is totally a gift from God who raises the dead.[30] The believers regard eternal life as real and true, sharing the very life of God himself.[31] Humans' present life is equally real and important because their life of obedience to God's will in this present life determines their destiny in the life to come.[32] This close relationship between the present and the future life is presented very vividly in the parable of the last judgment.[33] People who live a life of disobedience will suffer eternal

---

24. Mk. 5:23; Jn. 4:47 ff; Mk. 5:35 ff; Lk. 7:11 f:Jn. 11.1 ff.
25. Matt. 4:4; cf. Deuteronomy, a Book of Bible, (Hereafter, Deut.), 8:3.
26. Matt. 6:25 ff.; Lk. 12:15.
27. Jn. 5:21; cf. 11:25; 14:6.
28. Mk. 10:30; 1st Letter of St. Paul to Timothy, a Book of Bible, (Hereafter, I Tim.), 4:8.
29. Matt. 19:16; par. Mk. 10:17, Lk. 18:18; Matt. 25:46; cf. 2nd Letter of St. Paul to Timothy, a Book of Bible. (Hereafter, 2 Tim.), 1:10.
30. Matt. 22:31 f.; par. Mk. 12:26 f.; Lk 20:36 f.
31. Matt. 18:18; Mk. 9:43, 45.
32. Matt. 19:16; par. Mk. 10:17, Lk. 18:18; Lk. 10:25; Matt. 7:13; f.; cf. Lk. 13:23 f.
33. Matt 25:31 ff.

punishment, and the righteous and obedient will enter into eternal blissful life.[34] According to Christian understanding of life, it is not confined to historical time. The life lived here on earth in space and time points forward to eternal life which was promised by risen Christ to followers.[35] The human personality is not lost in the eternal life.[36] In this new form of life in heaven, each person will meet God and Christ face to face.[37] It will be a life of fullness, of righteousness, peace, joy,[38] and glory.[39] During their life in the world, Christians experience a healthy tension between their present lives here on earth and the future eternal life in heaven.[40] The life of the believer in the present existence is filled with a tension and conflict between the "old age" and the "new age," between the "desires of the flesh" and "those of the Spirit," and between the "powers of death" and "of life." These conflicting agents are at work in the believers and each of them are seeking to gain the mastery over them.[41] Through their belief in the Risen Christ and sharing of His new life, they are already aware of the promise of fullness of life which is not yet fully manifested.[42]

---

34. Matt 25:46; E. Schmitt, "*Life*," EBT II 499-503; V. Taylor, "*Life after Death I. The Modern Situation*," *Exp*T 76, 1964-65, 76-79; L Dewar, "*Spirit and Kingdom*," *Exp*T 82, 1970-71, 36-40.

35. 1 Cor. 15:26, 28; Rom. 6:22; Letter of St. Paul to Galatians, a Book of Bible, (Hereafter, Gal.), 6:8.

36. 1 Cor. 15:26, 28; Rom. 6.22; Gal. 6:8.

37. 1 Cor. 13:12; cf, 2 Cor. 5:7, 1st Letter of St. Paul to Thessalonians, a Book of Bible, (Hereafter, 1 Thess.), 4:17, 2 Cor. 5:8; Letter of St. Paul to Philippians, a Book of Bible. (Hereafter, Phil.). 1:23.

38. Rom. 14:17.

39. 2 Cor. 3:8 f.; Rom. 8:17.

40. Gal. 5:25.

41. Rom. 7:14-25; 8:10, 12 f.; Gal. 5:16 f.

42. Col. 3:3, 4; Rom. 8:18 ff; 1 Cor. 15:22; W.D. Davies, "Paul and the Dead Sea Scrolls: Flesh and Spirit," in K. Stendahl, ed., *The Scrolls and the New Testament*, 157-82.

Jesus is the source of divine life in this world as well as in the world to come because he is the true life.[43] Though the believers receive the life of God,[44] they are to express this experience of receiving God's life in their lives through a life of love[45] and joy.[46] Christians believe that eternal life with God will be a wholly perfect one. God himself will be with them face to face. There will be no more suffering, tears or death.[47]

Christian Perspective of Life and Resurrection Experience

The perspective of Life in the early Christian community was very much influenced and affected by the experience of resurrection of Jesus Christ by His followers. This post-resurrectional perspective of life is representatively presented by St. Paul and other early Christian writers.[48] They express that the divine life has power over death and all humans will also experience resurrection and eternal life at the end of this world by the power of the same divine life.[49] The believers experience Jesus Christ as the very embodiment of God's life-giving energy, conquering and defeating death and raising the dead to new life.[50] Human life has meaning only in Christ's

---

43. 1 Jn. 5:20;Jn 6:35, 48, 68; 10:28; 11:25; 12:50; 14:6; 17:2.
44. 1 Jn. 5:12; Jn. 6:40, 47.
45. Jn. 15-9-17; 1 Jn. 3:14.
46. Jn. 16:20-24; R. Bultmann, *Theology of the New Testament*, I, 1952-203-10, 345-52; *Primitive Christianity in its Historical Setting*, 1956; *The Gospel of John*, 1971; J.C. Coetzee, Life in John's Writings and the Qumran Scrolls, *Neotestamentica* 6, 1972, 48-66; C.H. Dodd, *The Interpretation of the Fourth Gospel*, 1953, 144-50.
47. Rev. 21:3 f; G.R. Beasley-Murray, "The Contribution of the Book of Revelation to the Christian Belief" in *Immortality*, *SJT* 27, 1974, 76-93.
48. 1 Cor. 15:4.
49. Rom. 14:9.
50. 2 Cor. 13:4.

everlasting life and this life goes beyond the grave. Jesus Christ is the author of a new life for humankind.[51] The life of Christians is not their own, but that of Christ: Christ lives in them[52] and they live the life of Christ.[53] Their life is justified by Christ[54] and by his life they will experience salvation and wholesome happiness in this life and hereafter.[55] Life of a human person is thinkable only in a body.[56] Life on earth is possible only through a physical body, and post-resurrectional life or life at the end of this physical world will be lived in a spiritual body.[57]

Christian Perspective of Life: A Life Guided by the Holy Spirit

The Christians also experience life in the Holy Spirit, who is a post-resurrectional gift to his followers by risen Christ.[58] Under the influence of the Holy Spirit, their life is radically changed. Though the Christians experience power of the new life they do not try to escape from struggle of living their everyday life in this world. On the contrary, the Christians are to serve their fellow-humans in Christian love with responsibility in the historical and cultural situation in which they live. In fact the believers are supposed to live primarily for others and

---

51. Rom. 5:12 ff.; 1 Cor. 15:20 ff.
52. Gal. 2:20; Phil. 1:21.
53. 2 Cor. 4:10.
54. Rom. 5:18.
55. Rom. 5:10; Phil. 2:16; cf. 2 Tim. 1:10; G. Bornkamm, Baptism and New Life in Paul (Romans 6), *Early Christian Experience,* 1969, 71-86; S.G.F. Brandon, Life after Death IV.F.F. Bruce, Paul on Immortality, SJT 24, 1971, 457-72.
56. 2 Cor. 5:1-10.
57. 1 Cor. 15:44.
58. Rom. 6:4.

not for themselves.[59] Since their life is surrendered to God[60] and to Christ[61] their life also shall show positive and tangible results[62] as they follow in the footsteps of Christ and serve the humanity even at the risk of suffering.[63]

For a Christian, life is not something static, but is ever growing and developing towards its fullness.[64] Receiving of the Holy Spirit through faith, marks the beginning of the fullness of Christian life.[65] Through the Holy Spirit, the Christian shares the life of Son of God and shares in his sonship, ever growing towards completeness of human life.[66] The Spirit is a divine power who has great impact upon the entire human life which is discernible by its effects.[67] The receiver of the Spirit experiences overwhelming divine love[68] and liberation from the deadening effect of the law and of sin's power.[69] The Spirit gives power to bring in moral transformation in human life with a vision for life.[70] The gift of the Spirit, in this way, is the beginning of a life-long process of the human life being saved, sanctified and being conformed

---

59. Rom. 14:7; 2 Cor. 5:15.
60. Rom. 6:10f.
61. Rom. 14:8; 2 Cor. 5:15.
62. Gal. 5:25, 26.
63. 2 Cor. 4:9 f.; 6:8 f.; Rom. 13:8-10; 14:11 ff.
64. Eph. 4:13-15.
65. Gal. 3:2 f.; Rom. 8:9; cf. 1 Cor. 2:12; 2 Cor. 11:4; 1 Thess. 4:8.
66. Rom. 8:14-16; Gal. 4:6; E. käsemann, "The Righteousness of God" in Paul, *New Testament Questions of Today*, 1969, 168-82.
67. 1 Cor. 1:4-7; Gal. 3:5.
68. Rom. 5:5.
69. Rom. 8:2; 2 Cor. 3:17.
70. 1 Cor. 6:9-11.

to the image of Christ,[71] which is the goal of life for a believing Christian.[72]

The life of believers in Christ ultimately belongs to the new age, wholly like Christ, wholly under the Spirit's direction.[73] The lives of the believers are based on this hope that the Spirit's activity in the present age will lead them to fulfilment of the promises and complete victory over the evil.[74] The believers have the responsibility to live their lives in the power of the Spirit. It means that their character be moulded by God according to the pattern of Jesus Christ. The believer's faith-life or good life is not something he/she achieves for himself/herself through his/her "good works." It is something which by attentive openness to God he/she allows the Spirit to produce through him/her, which is the "fruit of the Spirit."[75]

A Christian's daily life is lived by the Spirit, led by the Spirit, ordered by the Spirit.[76] Life guided by the Spirit is a shared gift. It is not lived for oneself alone. The Spirit is a centripetal force drawing the life of the believers together into the one body of Christ. A life of fellowship by the Christians arises out of the shared experience of the Spirit.[77] Through the influence of the Spirit, the believer experiences transformation in his life and his/her character is now changed

---

71. 2 Cor. 3:18; 2 Thess. 2:13; cf. Rom, 8: 28; 16:16; Gal. 6:8; Eph. 3:16 f.
72. W.D. Davies, "Paul and the Dead Sea Scrolls: Flesh and Spirit," in K. Stendahls, ed., *The Scrolls and the New Testament*, 157-82; J.D.G. Dunn, *Baptism in the Holy Spirit*, SBT second series 15, 1970; "2 *Corinthians* 3:17--The Lord is the Spirit," JTS 21, 1970, 309-20.
73. Rom. 8:11, 23; 1 Cor. 15:44-9; 2 Cor. 5:1-5; Eph. 1:14.
74. Rom. 5:5; 8:18-25; 15:13; Gal. 5:5; Eph. 1:17 f.
75. 2 Cor. 3:18, 6:6; Gal. 5:18-23, 6:1; Rom. 8:28; 9:1; 14:17; 15:13, 30; Col. 1:8.
76. Rom. 8:4-6, 13-14; Gal. 5:16, 18, 25; 6:8.
77. 1 Cor. 12:13; 2 Cor. 13:14; Eph. 4:3 f., Phil. 1:27; 2:1.

into the character of Christ.[78] The new life of the Spirit is a radical transformation that is considered by the Christians as a birth from above,[79] a new creation.[80]

Life of Faith, Love and Hope

Life of a Christian is lived in the Spiritual gifts of faith, hope and love.[81] These are not virtues practised in the life of a Christian, but they form the basis of the life of the Christian Community. The life of love, which a Christian is supposed to live, comes to fulfilment through a life of faith and hope. The new life of a Christian is determined by faith and hope and love.[82] Life, according to a Christian is meant to love and this commitment to love is only possible if one is committed in faith and has hope for an eternal life.[83] Life for a Christian is not merely an endless struggle for self perfection, but is something to be lived here fully, sharing with other fellow humans. The eschatological perfection is only the wholeness of this life which a person is given and promised.

Life for a Christian is not merely something that is lived for one's enjoyment and self satisfaction. It is also something lived for others as a sacrifice, as Christ's own life.[84] Sacrifice of

---

78. 2 Cor. 3.18 Rom. 8:9; Gal. 4:6; Phil. 1:19; cf. 2 Thess. 2:8.
79. Jn. 3:5-8; 1 Jn. 3:9.
80. Jn. 20:22; C.K. Barrett, "The Holy Spirit in the Fourth Gospel" *JTS* New Series 1, 1950, 1-15 ; R.E. Brown, "The Paraclete in the Fourth Gospel," *NTS* 13, 1966-67, 113-32; L.Dewar "1 Cor. 15:45–Last Adam, Life-giving Spirit," in B. Lindars and S.S. Smalley, eds., 127-41; *Jesus and the Spirit: A Study of the Religious and Charismatic Experience of Jesus and the First Christians as Reflected in the New Testament*, 1975; and *Unity and Diversity in the New Testament: An Inquiry into the Character of Earliest Christianity*, 1977.
81. 1 Thess. 3:5, 8; 1 Cor. 13:13; Col. 1:4 f.
82. 1 Cor. 13:13.
83. Tit. 1:13; 2:2.
84. Rom. 12:1.

one's own life means a total commitment to the following of Christ in response to the personal experience of Christ. Christian life itself is the true sacrifice. Ritual sacrifices have less importance in the life of a Christian. Christ's love for humans and his death for the humankind is a sacrifice in love. Similarly, Christian's life of love becomes a fragrant offering and sacrifice when he/she lives his/her life in love as Christ loved the humankind.[85] The Christian community, body of Christ, is built up by life of love by his followers.[86]

Life is Maintained and Preserved by Christ and his Word

Life of a believing Christian is not maintained merely by natural bread alone. Christ and his Word is the true and living bread that keeps him/her truly alive and loving person.[87] His/her life becomes part and parcel of the Body of Christ.[88] Life of a Christian attains fullness and becomes meaningful in this world when he/she is able to live together with other believers in love sharing each other's life.[89] Life of a Christian is basically communication rooted in love.[90] Christian life is a life committed in love and service of the others letting others come into his life. It is this new perspective of life that was preached by Jesus Christ.[91] Without love, no fellowship or shared life is possible.[92] The Life of the Christian community as the body of Christ is built up by love.[93] In the Christian perspective of life,

---

85. Eph. 5:2.
86. 1 Cor. 8:1; 2 Thess. 1:3; Phil. 2:1 f.; Eph. 4:16; Col. 2:2.
87. Jn. 6:49-51.
88. Rom. 12:1 ff.; 1 Cor. 12:12-27.
89. Cor. 14:4-12. 19, 23, 28.
90. Acts. 2:42.
91. Matt. 5:43-47; 19:19; 22:37-39; Mk. 12: 30-33; Lk. 6.27-35; 10:27; Jn. 13:34-35; 14:15-21; 15:9-17.
92. 1 Cor. 14:1; 16:14; Eph. 1:15; 3:17 ff.
93. 1 Cor. 8:1; 2 Thess. 1:3; phil. 2:1 f.; Eph. 4:16; Col. 2:2.

it is often blessed and taken care of by God continually, especially at every climatic point of life such as at birth,[94] at marriage[95] and at death.[96]

Wholesome Freedom Essential for Human Life

Life of a believer in Christ is one of freedom, free to love, free from the clutches of human-made laws that stifle human life, free from superstitions and free from material forces.[97] It is a new way of life in wholesome freedom according to the will and plan of God. When humans live their lives according to the will and plan of God, all the human-made good laws are included in it. This new-found freedom comes from his life with God,[98] fellowship with God[99] and with one another. Life of the believer is not according to the flesh but according to the Spirit.[100] The Holy Spirit works in the believer, becoming the principle of his life.[101] Those who belong to him have life and have it abundantly.[102]

Believers have come to fullness of life in Jesus Christ.[103] Thus Christ's joy becomes their joy.[104] This joy characterizes

---

94. Ruth (A Book of Bible), 4:13f.; Rf. Lk. 2:34.
95. Genesis, a Book of Bible, (Hereafter, Gen.), 24:60.
96. Gen. 48:1 ff.; 49:28 ff.
97. Gal. 5:1; Letter of St. James, a Book of Bible, (Hereafter, Jas.), 1:25; 2:12.
98. Rom. 6:22; Gal. 5:1. 13; 1st Letter of St. Peter, a Book of Bible, (Hereafter, 1 Pet.), 2:16.
99. Rom. 7:10; 10:5.
100. Rom. 8:12f.; Gal. 6:8.
101. 2 Cor. 3:17; Rom. 8:1 ff.; Gal. 5:18.
102. Jn. 10:10.
103. Col. 2:10.
104. Jn. 15:11; 16:24; cf. 1 Jn. 1:4.

the life of the disciples in their walk with Jesus; it becomes complete.[105]

Brevity of Human Life

Christians know the brevity of human life.[106] Just like the life of grass is short, similarly human life is very short, but God takes care of human life.[107] A Christian also knows that humans cannot influence the physical age or span of life they attain; it is a gift of the Creator God.[108] With all their worrying, humans are not even able to lengthen their lives by a trifling measure of time.

Christian Moral Perspective of Human Life

Human life is created in the image of God's own life. This means that all types of immoralities, passion, evil desire, impurity, greed, lawlessness and covetousness must be strictly kept out of human life.[109] All these are worshipping other gods or practicing idolatry. The physical human nature enslaved in sin must be subjected to the rule of the new spiritual life, with its transforming and liberating power revealed in Christ.[110] It is by his/her way of life he/she will show himself/herself to be a follower of Jesus Christ.[111]

In the Christian perspective of life, money and material things have very limited place.[112] Human life is not just meant

---

105. Jn. 3:29; 15:11; 16:24; 17:13; 1 Jn. 1:4; 2 Jn. 1:12.
106. Job, a Book of Bible, 13:25; pss. 37:2; 102:4. 11; 103:15; Is. 37:27; 40:6, 7.
107. Matt. 6:30; Lk. 12:28; Matt. 6:30; Lk. 12:28.
108. Matt. 6:27 and Lk. 12:25.
109. Eph. 5:5; Col. 3:5; 2nd Letter of Peter, a Book of Bible, (Hereafter, 2 Pet.), 2:3, 14;2 Cor. 6:14.
110. Matt. 5:29f.; Col. 3:3 f.
111. Phil. 2:8 with 3:9.
112. Mk. 6:8; Lk. 9:3; 3:14; Mk. 10:21; cf. Lk. 18:22.

for food and drink,[113] but is here to live and experience God's reign and his righteousness on earth.[114] Selfish amassing of material possessions has its prototype in human's grasping after the forbidden fruit in the Garden of Eden.[115] It is an indication that life, limited as it is by time, circumstance and the vital interests of others, is no longer accepted thankfully from the hand of God. Love of money erects a selfish dividing wall against God and our neighbours. It drives a person who is possessed by it into utter isolation. Thus striving after wealth is the germ of total alienation from God.[116] Christian life after the pattern of Christ's life has to be righteous and wholesome dying to all sinfulness in this world. In Christian perspective both legalism and lawlessness are "fleshly" in so far as they both hold out a false promise of life on the basis of human's own efforts. Thus jealousy, for example, is a by-product of seeking life in popularity; ill-temper arises from frustration in seeking life in comfort and quiet. Life is something over which he has no ultimate control. But an unbelieving human has a tendency seeking to find life through his own works, thinking that life is in his control.[117] But a believer thinks that for such a person who tries to find life on the basis of his works, death becomes a present reality.[118]

True life is, therefore, the life which comes by God's grace, and is free from the obligation to seek autonomously one's own fulfilment. It begins with faith in Jesus Christ and loving others like Christ did. "Truly, truly, I say to you, he who hears my word and believes Him who sent me has eternal life; he does not come into judgment, but has passed from

---

113. Matt. 6:25-34; Lk. 12:11-29.
114. Matt. 6:33.
115. Gen. Ch.3.
116. Matt. 6:24; 27:5-6; 28: 12-15; Lk. 15:12; 16:13-16; Jn. 12:6.
117. 1 Cor. 15:56.
118. Rom. 7:9 f.

death to life."[119] "We know that we have passed out of death into life, because we love the brethren."[120] A follower of Jesus, as a disciple, is expected to live his whole life as an unconditional sacrifice for the humanity.[121] Sacrifice of possessions and a life of discipleship in poverty were demanded of the rich young ruler as the precondition of eternal life.[122] In Christian view of life, not only possessions but also family ties are less important in view of a life of discipleship of Christ.[123] His life is determined by the commandments of God and not so much by self-control in the sense of an autonomous ethic. His life in this world is guided by his faith in Jesus Christ. Human life in physical body is lived on this earth, which is created by God.

The negative Christian attitudes to life[124] are influenced by some of the writings in the Bible. There is nothing good dwelling within me that is in my flesh.[125] For humans, indeed, desire to achieve what is good, i.e. the true life, but actually they achieve the evil which they do not desire, namely death, because they do not allow God to care for them, but desire to care for themselves.[126] Human life is sometimes described as futile and vain.[127] Some writings are extremely sceptical about the meaningfulness of human life. They say all is "vanity."[128]

---

119. Jn. 5:24.
120. Jn. 3:14.
121. Matt. 10:24 f.; Jn. 11:16.
122. Matt. 19:16-22; Lk. 18:18-24.
123. Matt. 10:37 ff.; Lk. 14:25-33.
124. Rom. 8:20-22.
125. Rom. 7:18.
126. Rom. 7:25b.
127. Psalms, a Book of Bible, (Hereafter, Ps.), 103:14 ff.; Ps. 94:11; Job 14:1 ff.
128. Ecclesiastics, a Book of Bible, (Hereafter, Eccles.), 1:2; 2:1.

The writer finds no meaning in human life at all. Human life is merely a tragic existence from which one cannot escape and therefore remains resigned to tragic existence. This kind of perspective of life has come to the Christian thinking from historical experience of the absence of God's intervention in the midst of natural calamities and sufferings caused by injustice, exploitation and inhuman acts of the wicked.[129]

Goal of Christian Life

Christian view of life holds that life is not something to be just preserved, but is to be spent for others. It is the conviction among the believers that whoever saves his/her life will lose it; and whoever loses his/her life for a true cause and for the good news of Jesus will save it.[130] The believer's life should be lived for Christ and for his community. The goal of life, from the Christian perspective, is to be described in terms of righteousness.[131] Righteousness takes its character from God himself[132] and salvation is grounded in righteousness[133] which finds its expression in right behaviour.[134]

Conclusion

In this short article I have tried to present Christian perspective of life based on the Biblical writings of the early Church and Christian tradition based on these writings. From what is said above it is clear that Christian perspective of life is very much positive and it gives importance both to the physical and

---

129. Eccles. 2:25; 3:13; 7:14; 9:7 ff.; Jeremiah, a Book of Bible, 2:30; Isaiah, a Book of Bible, 49:4; Malachi, a Book of Bible, 3:14.
130. Mk. 8:35; Matt. 16:25. Lk. 9:24; cf. Jn. 12:25.
131. 2 Pet. 3:13.
132. 1 Pet. 2:23; Revelations, a Book of Bible, 16: 5.
133. 1 Pet. 3:18; 2:24 ff.
134. 1 Pet. 2:24; 3:12; 4:18.

spiritual aspects of human life. Human life, as we have seen is precious, beautiful and goal-oriented gift by the Creator God. If humans orient their lives to the right goal, they will be wholesomely happy in their lives in this world and will continue to live eternally in the presence of God, sharing his own life in the world to come.

References

A Greek-English Lexicon revised by H.S. Jones, 1940.

Amdt, W.F. and Gingrich F.W. (1957). *A Greek-English Lexicon of the New Testament and Other Early Christian Literature.*

Barrett, C.K. (1947). *The Holy Spirit and the Gospel Tradition.*

Bruner, F.D. (1970). *A Theology of the Holy Spirit.*

Bultmann, R. (1952). *Theology of the New Testament,* vol. I.

Button, (1918). *Spirit, Soul and Flesh.*

Dewar, L. (1959). *The Holy Spirit and Modern Thought.*

Dodd, C.H. (1953). *The Interpretation of the Fourth Gospel.*

Fison, J.E. (1950). *The Blessing of the Holy Spirit.*

Ladd, (1975). *A Theology of the New Testament.*

Smally, S.S. (1975). *Jesus and the Spirit: A study of Religious and Charismatic Experience of Jesus and the First Christians as Reflected in the New Testament.*

Von Rad, G. (1966). "Righteousness and Life in the Cultic Language of the Psalms, in *The Problem of the Hexateuch and Other Essays.*

8

Religious Concept of Life

A.S. Inam Shastri

The Concept of life is a very complicated matter to discuss except in the light of religious view points. Life is nothing but a period of preparation for death. The objective of life is death. Life is mortal and mortality is for immortal life which is death. There is no consideration and understanding of life if there is no sense of death. However, it is not easy to understand the fact of life as the people live their lives as per their belief, their religion, their socio-cultural background, their faith as well as their environmental situation. The importance of life is always based on the facts of the individual's mind and sense. Someone pays a lot of importance to his life while another considers it a waste of time. There are a few who enjoy their life, and others want to get rid of it.

The life is a combination of soul and body. There is no life in the absence of any one of the two. If these two will be separated there will be no existence of life but death. The only important factor is that no one can rejoin these two things, because one is observable and the other one invisible and abstract.

The life is God-gifted. It is originated by God. The concept of life may not be well understood without taking the religious view-point into consideration.

According to Christian view-point "The humans are created by God and they have a responsibility to accept his reign. The Old Testament's understanding of creation of humans and the human society is expressed in the Book of Genesis. It says that men and women are created by God in his own image and likeness (Gen. 1.26-27 5.1-2). The synoptic Gospels uphold this belief (Mt. 19.4, Mk.10.6). As the image of the Creator God, humans cannot be reduced to mere dust even after death.[1]

The Psalmist expresses his faith in Yahweh (Lord) that he will be rewarded for his righteousness differently from the wicked. If at death all are reduced to dust, then this faith has no relevance.[2]

The synoptic Gospels hold that humans are different from animals and they are composed of body and soul. The human body can be destroyed like that of an animal, but his soul cannot be destroyed in this world (Mt.10.28). Both the human body and human life are more precious than anything in this world (Mt 6.25; Lk.12.22) and, therefore, one should not endanger their precious life for the sake of things in this world (Mt.16.26, Mk. 8.36, Lk. 9.25). Humans, as creature of God, are totally dependent on Him as little children on their loving father (Lk. 12.47, 11-12;16-31). Their life and death depend on His will (Lk. 12.20). Jesus, therefore, invites the humans to cast all anxieties away and totally depend on the love and care of God, the Father, and seek his Kingdom (Lk.12.22-31 Mt 6.25-33) and He will give it to them (Lk. 12.32).

According to *Bhagavad-Gītā* (Concept of Universe):

---

1. J. Pathrapankol, (1922). "Structure of Hebres Language," in *Unique and Universal*, ed. J.B. Chethimettan, Bangalore: Dharmaram College, 65.
2. J.L. Mekenzie, (1972). "Aspects of old Testament Theology" in the *Jerome Biblical Commentary*, vol. 2 ed. R.E. Brown and others, Bangalore: T.P.I.

> I am the origin of all; from me all (the whole creation) proceeds, knowing this, the wise worship me endowed with conviction. (*BG* 10.8)
>
> Know that all beings have their birth in this (my nature). I am the origin of all this world and its dissolution as well. (*BG* 7.6).

In the above quoted texts the *Bhagavad-Gītā* speaks of the material world and all the living things in it as evolved from the Supreme Lord, as the spider produces a web from its own body. This universe will be ultimately dissolved into himself (cf. *BG* 2.17, 6.30, 7.12, 19; 8.18-22, 9.4-17; 10.8-20; 11.13-15; 13.27-28,14.3-4; 15.14). God is manifested in and through the beings on earth which are originated from him. Accordingly, the origin of the universe is not "out of nothing" but is a part of the Supreme God. The Universe and all things in it are also sustained by him through continued action (*BG* 3.22-24; 7.5, 7.8-11).

The Hindu tradition from the ancient times has maintained that humans are different from other living beings, not so much in their relation to God. The oldest Indian definition of human person derives from the Brāhmaṇas of late Vedic period, which says, "Man is the only animal that offers sacrifices."[3]

In certain passages the idea is amplified regarding the life as it is highlighted in Holy Quran Sura VI (32) Anam:

> *What is the life of this world*
> *But play and amusement?*
> *But best is the Home*

---

3. H. Von Stieten Cron, (1887). "Man and Salvation in Hindu Religions: Hindu Perspectives" in *Christianity and worlds religion;* ed. H. King, London: Fount Piper Bachs — 212.

In the hereafter, for those
who are righteous.
Will ye not then understand.

Further classification is available in the same Sura:

The life of this world is but empty;
What is serious is the life hereafter,
The teacher of God's Truth is not baulked
By frivolous objections or insults
or persecution. The wicked will be
Cut off to the last remnant. God's wisdom
Pervades the whole of His Creation,
And in His hands are the keys of Unseen
And the secrets of all that we see.

LVII-20

Further in Sura. Hadid or Iron the life is considered as under:

Know ye (call), that
the life of this world
Is but play and amusement, 5302
Pomp and mutual boasting
and multiplying, (in rivalry)
Among yourselves, riches
And children.
Here is a similitude: 5303
How rain and growth
Which is bring forth, delight
(The hearts of) the tillers 5304
Soon it withers; thou
will see it grow yellow;
Then it becomes dry
And crumbles away.
But in the Hereafter
Is a penalty severe
(For the devotees of Wrong).
And Forgiveness from God

And his God Pleasure
(For the devotees of God)
And what is the life
of this world, but
Goods and chattels
of deception? 5305

5302

As per the above passage, in this life people not only play and amuse themselves and each other, but they show off, and boast, and pile up riches and manpower and influence, in rivalry with each other.

5303

The next passage of the above verses indicates that God's mercies are free and open to all, like his rain. But how do men make use of them? The good men take the real spiritual harvest and store the spiritual grain. The men who are in love with the ephemeral are delighted with the green of the trees and the grass; but such things give no real nourishment; they soon wither, become dry and crumble to pieces, like the worldly pleasures and pomps, boastings and tumultuous possessions and friends.

5304

"Further in the original verses of Quran "tiller" is used in the unusual sense of husband/man; because they sow the seed and cover it up with soil. But the ordinary meaning "Rejection of Truth" is not absent; the allegory refers to such men.

At last, it is finally and clearly indicated that many of the attractive vanities of this world are but nets set by the Evil one to deceive man. The only thing real and lasting is the God-given life lived in the Light of God.

XVII-33

Sura Cave (Kahf) indicated as under:
Nor take life-which God
Has made sacred-except
for just cause. And if
Any one is slain wrongfully,
We have given his heir
Authority (to demand Qiases
or to forgive) : but let him
Not exceed bounds in the matter
of taking life; for he
is helped (by the law).

2216

Under the strict limitations laid down there, a life may be taken for a life. The heir is given the right to demand the life; but he must not exceed due bounds, because he is helped by the law. Some commentators understand "he" in "he is helped (by the law)" "to refer to the heir of the person against whom Qiases is sought. He too will be helped by the law, if the heir of the first slain exceeds the bounds of law.

However, as per the Islamic point of view the life of this world is just a struggle if it is used for play and amusement and not for just and for the life-hereafter in the Light of God.

9

Law of Life: The Cosmic Principle

Baidyanath Saraswati

WHAT does a fish know of water in which it swims all its life? It knows, and knows not. What does a man know of Nature, the cosmic Tree, in which he is rooted? Does he know or know not? How does he live?

On the symbolic representation of the cosmic Tree, it is said: This Tree is the great World Mother, the Goddess of Nature who nourishes all life with the milk of her breasts. Hence the metaphor Fig Tree, with its milky juice, the most sacred tree known as the *aśvatha*, the *baṭa* or banyan, and the *ūdambara* of the ancient Indo-Aryans. The root of this Tree is said to be above and branches beneath, whereas all the other trees have their roots in the Earth and their branches in the Sky. There is a mystical tie between Tree and Life. Man resembles a tree growing from *Brahman*. One of the Upaniṣads expresses this idea vividly:

As a tree of the forest
Just so, surely, is man.
His hairs are leaves,
His skin the outer bark

From his skin blood,
Sap from the bark flows forth.
From him, when pierced, there comes forth
A stream, as from the tree when struck.

His pieces of flesh are under-layers of wood.
The fibre is muscle-like, strong.
The bones are the wood within.
The marrow is made resembling pith.

A tree, when it is felled grows up
From the root, more new again;
A mortal, when cut down by death-
From what root does he grow up?

Say not "from semen,"
For that is produced from the living,
As the tree, forsooth, springing from seed,
Clearly arises without having died.

If with its roots they should pull up
The tree, it would not come into being again.
A mortal when cut down by death-
From what root does he grow up?

When born, indeed, he is not born (again)
who would again beget him?
Brahman is knowledge, is bliss,
The final goal of the giver of offerings.
Of him, too, who stands still and knows it?

— *Bṛhadāraṇyaka Upaniṣad*

Indian tradition speaks of thought (or thinking about thought) as a transcendent phenomenon in the scheme of things. One of the earliest works on human organization is the *Mānav Dharma Śāstra*, literally the "Sacred Science of Man." This text is said to be the Word of *Manu*, the first Man, the first Life. It symbolizes the cosmic principle or Law of Life. *Manu* is the symbol of the Universal consciousness which exists in the transcendent, as also in the differentiated units of creation, or in each individual centre which is but a ray from its source, the universal sun. *Manu* is the Mind of the Creator, the principle of Ego. The Mind of the Creator is an ocean. There is, as sages have said, a complete identity between the Creator and the cosmos. Manu has a twofold aspect: he is above Time and Space, and he is also the presiding ruler of each cosmic cluster of Time. Each Manu in his incarnation is the sovereign ruler of his dynastic Time, and it is his charge that the primeval ordinance be fully and perfectly fulfilled in his period. It is the Manu-Principle that governs the traditional Indian vision and society. There is another principle which touches on a crucial problem that could easily be the starting point for understanding Life in the Universe. It is the Matter-Principle, the fundamental Elements in their gross and subtle aspects of the reality. The third is the Creating Principle, which is the cause and the ultimate basis of the world. These three interrelated principles are set in motion from the beginning of the world to descend into the sacred history of Life, and thereby into the present time. What we seek to indicate is that the Sacred Science proceeds from another source of knowledge, which is infinite, and that it is continually transmitted to this world to the extent to which man has not ceased to read the signs.

The Upaniṣadic sages had described *Brahman* as *bindu*, the dimensionless point, literally a drop. The *bindu* consists of the unity of two: one static and the other dynamic. It is the dynamic aspect that expresses itself as the Universe of multiplicity, but it essentially remains One, the epitome of all. In geometry the point is the Sun, i.e., the centre with a circumference. There is a centre and a periphery, diameter and radius. The circle is a point expanded: the meeting of four equal lines forms a square, and three lines come together to form a triangle. These primary forms are fundamental to Indian thought.

The concept of *śūnya*, zero, when applied to *Brahman* the Absolute, is unthinkable in itself. Metaphysically it is identified with both fullness and emptiness. The description of *Brahman* as full and empty shows that it is beyond the dialectical oppositions of time and timelessness, formed and unformed, mortal and immortal, actual and beyond; and so on. The whole Universe is regarded as a "living organism." The *Hiraṇyagarbha*, the Golden Egg, is a cosmic metaphor attributed to the origin of the Universe. Here gold symbolises the principle of light shining forth from within the womb of darkness. The sap within the Egg originally existed in an undifferentiated form but later were born the seven primeval Ancestors, or seven differentiated life forces (Mind, Life, and the Five Elements of matter). The Egg floated in Waters for a thousand years and then something triggered off. It split into two, and the whole Universe came into existence.

The organic units of the Universe are further clarified by an anthropic metaphor. From the limbs of *Puruṣa*, the Cosmic Man, everything originated: strophes and melodies, horses and every animal, the four human castes the Sky, the Atmosphere, the Ocean, the Earth, the Moon, the Sun, the Fire, the Wind, and the King of Gods. The idea of organic relationship is pursued to a high degree of sophistication. It recurs in the *Bṛhadāraṇyaka Upaniṣad*:

> *The self is the honey of all beings*
> *and all beings are honey for the Self.*

The mathematical description of *Brahman* as zero reveals what is infinite.

a-a=o; a+o=a; a-o=a; axo=o; a./o. becomes infinity.

The geometric description highlights the plurality of primary forms: the square, the circle and the triangle. In the description of the Cosmic Egg or Man, the One is not the all, that is, the One is not the sum total, the One subsists in the many. Although the world of plurality has emerged out of the One, the One is not affected by the defects of the many. Here, the One is to be understood as a dimension of the Self. The primordial oneness arises from within the Self, hence it makes differentiation possible. The individual Self is valuable precisely because all generic and ontic differences are included in the One, the universal Self.

Anthropologists' conception of Life is grounded in the theory of evolution. In biology, evolution refers to slow changes in the process of growth. Applied to human evolution, it implies that Man represents the pinnacle of evolution and his world is pinned to a ladder of progressive development.

There is another theory which looks inward upon Man and the Universe. Its basic assumption is that, like the rest of the material world, Man's body is made up of the Five Primordial Elements: (i) Earth, (ii) Water, (iii) Fire, (iv) Air, and (v) Sky. Each major organ is said to have its own builder. Head and ears are associated with Sky, neck and chest with Air, stomach with Fire, and body with Earth. The five fingers, from the little finger to the thumb, are associated with Earth, Water, Fire, Air and Sky, respectively. Elements are also known to have caused the origin of bio-social types. For example, ironsmiths relate themselves with Earth, carpenters with Water, coppersmiths with Fire, sculptors with Air, and goldsmiths with Sky.

The theory of the Five Elements leaps forward to the vision of the Five "sheaths" or envelopes: (i) the gross body or the vegetative envelope; (ii) the vital body; (iii) the mental body; (iv) the poetic or higher intellectual envelope; and (v) the beatific envelope, which is the highest in the realization of self-knowledge and self-control.

Man's relationship with the Universe is explained by a powerful imagery of the Cosmic Man, with a thousand heads, a thousand eyes and a thousand feet, encompassing the planet Earth on all sides, exceeding it by ten fingers' breath. This does not make the world anthropocentric. The Cosmic Man is neither being nor non-being. From him the shining Cosmic Egg was born, and from this again He himself emerged. In other words, He became the creature of his own creation. From this creature the Four human castes (*varṇa*) took birth. All kinds of life emanated from the cosmic body, each invested with Cosmic powers and functions.

References

Agrawala, Vasudeva, S. (1968). *Matsya Puraṇa: A Study*, Varanasi: All India Kashiraj Trust.

Hume, Robert Ernest (1984). *The Thirteen Principal Upaniṣada*, Delhi: Oxford University Press.

Saraswati, Baidyanath (2001). Alternative Paradigms in Anthropology: the Cosmic Anthropology Principle in *The Nature of Man and Culture* (ed) Baidyanath Saraswati, New Delhi: IGNCA and Aryan Books.

Part II
Science and Society

10

Technology and Destiny of Man
What Can We Learn from the History of Life?

Pratul Kumar Saraswati

CONCEPTUAL thought was an adaptive breakthrough in the evolution of man. It led to technological development which we believe is our unique possession. Within a few thousand years of its development it has started showing signs of disastrous consequences not only on its mentor species *Homo Sapiens* but on the entire living system. The conscious environmentalists' advice towards mitigating the disaster is anthropocentric and at best aims to postpone it to the next generation. Stopping technology is not a viable proposition. Important is the type of technological development. Considering that technology is the result of conceptual thought in man, and that conceptual thought is one of the traits in evolution, it is logical to assume that the processes behind a successful technology would be compatible with those determining the successful history of life. Based on this premise some essential features of technology can be examined to create a friendly environment for the survival of life on this planet.

The value of technology lies in the impact that it has on shaping society. In Toffler's[1] words, today's technology is

1. A. Toffler, (1971). *Future Shock*, London Pan Books,: p. 517.

creating "Modular Man" — a disposable person. It fosters impermanence, forming a throw-away society in which not only are material things temporary but even human relationships are becoming increasingly short-lived. Family is becoming obsolete and the old and elders are loosing their importance in society. Man is the victim of the pace of technological change as well. The fast changing technology is creating an overwhelming choice of material commodities and cultures which he may not be prepared to cope with.

Is it merely a coincidence that the impact of technology that we observe today is against the very novelties of our evolution? The unique characteristics of man that it inherited through evolution include a slow rate of development and a long "post-maturity" period. A slow rate of development prepares it to acquire greater human skills. If so, how can the fast change in technology be assumed to be advantageous to man? Equally important is the utility of the individuals who have crossed the reproductive stage and are in "post-maturity" period. This period is longer in human beings compared to any other animal and has functional importance in the life of the species. When this trait of human evolution is so important, can the technology creating Modular Man be called progressive? Making the old useless to society, in Huxley's[2] words is "to rob man one of his unique characteristics whereby he utilizes tradition to the best advantage."

The patterns of evolution of life during the last six hundred million years are suggestive of the survival of simple forms. The complex forms have perished whereas simple forms survived the great extinctions to carry on the life cycle. The fossil records in the earth's crust are full of such examples. The ammonites developed complex sutures and bizarre shells in Cretaceous times, shortly before their extinction. Its close

---

2. J. Huxley, (1951). *Man in the Modern World,* New York Mentor Book,: p. 191.

relative, the simple *Nautilus*, continues to occur in the Pacific waters. Many complex offshoots developed from the unicellular forminifera *Globigerina* at various times in the past 150 million years of its existence and became extinct. This simple form itself continues and would probably continue to be the progenitor of many divine life-forms in future. But what does today's technology do? Its catchword is to meet the needs and choice of individual man. In this process it will produce complex individuals with very specific requirements and thus promote a society where no two individuals will probably be similar. Too much of diversity, as the modern technology is creating, is not a favourable feature. Some extinct group of animals such as trilobites in the Cambrian time and ammonites in Cretaceous showed a "boom and burst" pattern of diversity.[3] A high extinction rate in them is accompanied by high origination rates (and thereby a high diversity). Evolution of life is also benefited by stable environments as not many species can tolerate fluctuating conditions. A highly variable and fast changing technological environment will create similar effect which not many would be able to cope up with.

Continuity from simple to advanced technology is important. A high technology thrusted on a primitive society will break the adaptability of man. The great extinctions in the history of life, when a large group of organisms were wiped out from the earth, were accompanied by major regressions of the sea. The regressions broke the continuity of the marine environment. The continental shelves, which housed a wide variety of marine fauna, reduced in area and obviously the marine forms could not be expected to adapt soon to terrestrial condition.

---

3. K.W. Fless, Erben, H.K. *et al.* (1986). Causes and Consequence of Extinction. In, Raup, D.M. and Jablonski, D. (eds.) *Patterns and Processes in the History of Life*, Berlin Springer Verlag,: pp. 235-57.

A technology changing at slower pace, maintaining an optimum diversity and keeping continuity appears natural and ideal. Developing technology conforming to these ideals can prolong our existence on this planet. Any suggestion about the eternity of *Homo Sapiens* is a wishful thinking. After all, it is one among the several thousands of species that have appeared and got extinct in this drama of evolution of life. Its extinction will not be mourned more sympathetically than the deaths of nearly five thousand animal species at the verge of extinction today.

Note

This essay follows my reading of Alvin Toffler's famous book — *Future Shock*. I have tried to give an alternative explanation to his concern about the consequences of fast changing modern technology. For freely using the sociological jargons from his book, I acknoweldge him.

11

Life: A Scientific View Point

Surendra Nath Mishra

Introduction

THE Earth was formed some 48,000 million years ago. During the first 15,000 million years organic compounds are believed to have been formed from atmospheric components such as $H_{2,}$ $NH_{2,}$ H_2O and methane under activation by ultraviolet light, lightning, shock waves, heat and other forms of energy. Among these compounds were some 30 or 40 primordial biomolecules including amino acids, purines, pyrimidines, pentoses, hexoses, fatty acids, and others. The subsequent chemical evolution of organic molecule probably proceeded in a stepwise manner. Primitive cell-like structure may have formed by the process of conservation or by micelle formation driven by hydrophobic interactions. The capacity for further evolution developed later on. Development of the genetic code may have the basic event in the origin of life. At the centre of the story is the process of self-organization of matter.

Living Matter Differs from Non-Living

Why does living matter differ so radically from non-living matter which also consists of several inanimate molecules? Why does the living organism appear to be more than the sum of its inanimate parts of its composition? No doubt this difference is nominated in one word "Life." What is life? What is its form and formula? Early thinkers and philosophers have tried to explain, and have answered that living organism is endowed with a mysterious and divine life-force.This doctrine is called vitalism. But later this thought was nullified and rejected by modern science.

Today the central goal of the science of bio-chemistry is to determine how the collection of inanimate molecules, found in living organism, interact with each other to constitute, maintain and perpetuate the living state.

Identifying Characteristics of Living Matter

The most conspicuous characteristic of living organism is that they are complicated but highly organised. They are composed of cells which possess complicated internal structures containing many kinds of complex molecules, but the inanimate non-living matter around us like soil, water and rocks usually consists of simple compounds with little structural organisation. The small character of living organism is that each component part of living organism has a specific function. The third and most important attribute of living organism is their capacity to extract and transform energy from their environment, which they use to build and maintain their own structure from simple raw material, and also they can carry out other forms of purposeful work. Non-living matter does not have the property to utilize external energy to maintain its own organization. Usually they decay when they absorb external energy in the form of heat or light.

The most extraordinary property of living organism is their capacity for self-replication through generation after generation.

Bio-chemistry of Living State

Bio-chemistry has achieved success in explaining many cellular phenomena. Scientists have come to the conclusion that biology is chemistry. Today it is assumed, as a working philosophy, that all biological phenomena are ultimately molecular in basis.

This is because the molecules found in living organism not only conform to all the familiar physical and chemical principles governing the behaviour of all molecules, but in addition they interact with each other in accordance with another set of principles that we shall refer to as the molecular logic of the living state. They should be regarded as a set of ground rules that govern the nature, function and interaction of the specific types of molecules found in living organisms. Not all the principles comprised by molecular logic of living cell have yet been identified. Some are only dimly perceived and not yet provable.

Energy Transformation in Living Organism

The molecular complexity and the orderliness of structure is the basic character of living organism, performing most of the physical and chemical processes. Second law of thermodynamics states that all the physical and chemical processes always process with an increase in the disorder or randomness i.e. its entropy.

Living organism do not observe exception to the law. Living organisms thus cannot consume or use up energy; they can only transform one form of energy into another. They absorb from their environment a form of energy that is useful to them under the special condition of temperature and pressure in which they live and then return to the environment an equivalent amount of energy in some other, less useful form. This way we can say that living organisms create and maintain their essential orderliness at the expense of their environment. In the language of thermodynamics living organisms are open system because they exchange both energy and matter with the environment. Living cells are highly efficient in handling energy and matter. They greatly exceed most man-made machines in the efficiency with which they convert input energy into work performed. Living cells function as isothermal chemical engines. The energy that cells absorb from their environment is transformed into chemical energy, which is then used to carry out the chemical work, osmotic work and the mechanical work.

In our survey of bio-chemistry we have many glimpses into the group of organizing principles which we have called the molecular logic of living matter. From the view-point of a physical scientist, cell may be regarded as complex collection of organic molecule which are self-organizing and self-replicating, which are capable of exchanging energy and matter with their environment by means of enzyme-catalyzed organic reaction and which appear to function on the basis of maximum economy of parts and processes. But it is still an unanswered question as to how did organic bio-molecules learn to interact with each other and organise themselves, or how life arises or what may be the real meaning of life. In spite of the latest knowledge of bio-chemistry we have acquired at this movement, we cannot define "Life" accurately enough and determine at what point it arose in the chain of events, leading from preformed macromolecular component to a complete cell. However, it is generally agreed that a minimum requirement for life is one or more informational macromolecules capable of directing their own replication. One must then ask: what is that which has primacy in the origin of life, proteins or nucleic acid? Probably nucleic acid and proteins had come together to form the first real precursor of a living cell. Capacity for self-replication and further evolution was the fundamental turning point in the formation of life. Development of the genetic code may have the key role at this event. At the centre of the problem is the process of the self-organization of matter.

References

Fox, S.W. (1973). *Origin of the Cell: "Experiments and Premises"* Naturwiss; 60: pp. 359-69.

Lehniger, A.L. (1975). *Biochemistry* IInd edn., pp. 3-14.

Oparin, A.I. (1965). The origin of life and origin of enzyme in *Adv. Enzymol* 27: pp. 347-80.

12

Life, Nature and the Sacred in Mesoamerican Cosmo Vision

Yolotl González Torres

IN the Western world, the separation of men from nature has increased with growing urbanisation and with the following withdrawal from it of the people who live in the big cities and have no contact with the countryside; they never see a night sky full of stars, nor do they know the plants or the animals, except the domestic cats and dogs. It is no exaggeration to say that some children have never seen a cow.

The separation from nature prevents the relationship which drove our ancestors to preserve, study, admire, fear and worship her, through different manifestations, and at the same time to imagine, practise and experience situations which lead them to a relationship with the sacred, and which produced symbols and models of Gods.

"Man dialogues with nature, he does not subdue her," say the Tojolaban, an ethnic group from Mexico.

Mesoamerica is a culture area, which covers most of Mexico and Guatemala, and shares a cosmo vision from ancient times to the present. There are now 75 ethnic groups which speak languages from the *utoaztecan*, the *macrotomangue* and the *Māyā*, and who form 14 linguistic stocks, constituting 20% of the population. Nahuatl was the most widely spoken language in this area, and the language which was spoken by the Aztecs who ruled most of Mesoamerica when the Spanish defeated them and imposed the Catholic religion, which has now mixed up with the ancient religions.

The rest of Mexicans, mostly *mestizo* of Spanish, Indians and African negroes, who were brought as slaves, share a sort of Western civilization. Let me say on the concept of life of the Mesoamerican people. Life in *Nahuatl* is called *Yoli,* from the root *Yol,* movement, *yollotl* — heart has the same roots, it is the core of life and movement.

Life is connected to earth, mother earth, from where all nourishment, which is life, comes. And the main nourishment for the Mesoamericans, and for most native Americans was maize.

The conception of the world about the relationship with nature between the monotheistic peoples and the Mesoamericans is quite different. For the first, God, a male God, made plants and animals to serve man, the only one who possessed a soul; while among the second not only animals but all nature had life and power. There is an aspect of the sacred in nature, therefore she is propitiated and deified in different degrees and forms.

This concept of the world could not be understood by Hernando Ruiz de Alarcon, one of the friars who came to convert the Indians, as we can read in text of the VXII c., who wrote that:

> don't say all, because I have not been throughout the land, but there must be little difference, as I understand, they are all very easy to persuade about what they want us to believe. So that, because of their ignorance, they had and have several Gods, and ways of worshipping them so different, that when I came to find the foundation, and what they all are, we find so little to take hold off, as if we could appreciate it in the smoke or the wind. What is certain is that most, or

> almost all, the present ways of worship or idolatryc actions that we now find and of what we can judge are the same, their ancestors used to have; they have their root and formal foundation to believe that the clouds are angels or it is so much the ignorance and simplicity of almost all the Indians, and God capable of being worshipped; and the same they judge about the winds, and so they believe that they live in all parts of earth, like the hills, mountains, valleys and cliffs. They believe the same of the rivers, lagoons and springs, and to all of them offer candles and incense, and what they venerate most, and almost all have for God is fire.

Practically all non-monotheistic peoples believe that nature and cosmos, in general, share a vital energy which brings them together and so they consider that it is necessary to maintain an equilibrium, a harmony between all forces that circulate and move the cosmos.

These forces or energies are conceived as a complementary dichotomy of a masculine and feminine energy, positive and negative in the physical and not moral sense, and which could be compared to the Chinese *yin* and *yang*, which we have equalled to the *Nahuatl* words *tona* and *celtic* or *ceua* with the same connotation. This first dichotomy of the universe of hot-dry and masculine, and cold-humid and feminine, both indispensable and with equal importance and value, must exist in the adequate proportion, so that there is an equilibrium and harmony in the Universe. Equilibrium, in a certain moment, means death, because in the Mesoamerican cosmo vision *death was indispensable for life to exist.*

Among the Nahua the maximum source of hot-positive energy was the Sun called Tonatiuh, the one that goes producing life or *tona*; but an excess of *tona* could bring a drought or even death. Which means that even if the Sun was indispensable for the existence of the living beings, excess of its heat could be harmful.

On the other hand, all beings in the Universe, including human, shared a part of *tona*. The amount and quality of *tona* in each human being depended on the day they were born. And this was contained in a ritual "calendar" or cycle, called the *tonapohualli*, the count of *tona*, which was registered in the *books of tona*, which were consulted by specialists called *tonapuhque*. This ritual cycle was formed by 13 signs which included objects like flint, or house, plants, like reed or flower, and animals like rabbit or jaguar, plus abstract signs like wind or movement. These 13 signs were combined with 20 numerals, giving 260 possible combinations, which gave the personality and the fate of each individual, very much working like astrology.

Even if these signs marked the destiny of Gods and human beings, it was not necessarily definitive, and there was the possibility to change it by performing some rituals.

The *tonapuhque* specialised in curing certain types of diseases related to excess or loss of *tona*, which was concentrated in the crown of the head, from where it escaped at the time of death.

There were also specialists in charge of handling the cold energy which was more related to earth and water. They were the *graniceros* "healers." The rituals of these people usually take place in hills and caves, and they have the faculty of calling the rains, and they cure by "raising the shadow." Some of these *graniceros* still live and perform their rituals, and are in communication with the two mountains which are very near Mexico city, one of which is an active volcano which is constantly throwing smoke, ashes and sometimes even lava. The people call him Popocatepetl "the smoking mountain," and the *graniceros* call him, Don Gregorio as it is supposed to be a male mountain, in opposition to the other mountain which is the "White Woman."

All these specialists in healing handle this dichotomy of cold and hot in the classification of diseases as well as in what is prescribed to the patients for their cure. As you must have noticed, this type of classification does not happen only in Mesoamerica, but is common to many peoples in the world.

The Mexica had a series of complicated ceremonies, throughout the solar year of 365 days, which was divided into 18 months of 20 days. The path of the Sun throughout the year was taken into consideration, marking the solstices and the equinoxes, which implied the bigger or lesser strength of the solar energy, which meant that it was necessary to perform rituals which would help the Sun in his annual run. In the same manner it was believed that if there was an excess of Sun and a threat of drought, the Gods of water, his counterpart had to be propitiated, so that they would water the earth, and life could come out of it.

These ritual cycles were, besides, united to the cycle of the growth of plants, especially maize.

Earth was considered a living being, the father, mother of every living being; it was conceived as an androginous monster, a sort of alligator which floated on the primordial waters, and which is described in the traditional Nahua texts with mouths in the joints of its limbs, and in the Maya iconography as a monster crocodile or dragon. It was also an aspect of the great creator god Itzamna.

Earth was the great nutritious mother, in which life germinates, from which plants, which feed the rest of living beings, grow and which at the same time receive all the bodies after the necessary death, so that the cycle of life continues.

The earth goddess was also imagined and represented by the Mexica in what I consider the most impactant image of feminine force in religious art of all times: Coatlicue, the one with a skirt of snakes, image in which one can feel the tremendum of the sacred feminine, which one could, in a way, compare to the great *śakti* of the Hindu cosmo vision. We think that this image must have been taken as a sort of *mandala*, a cosmogram, a symbolic map or guide of the telluric forces of the Universe which was done with the idea that the initiates could reach other states of consciousness.

There have been many interpretations of this high stone sculpture since Leon de Gama, in the XVI century, who interpreted her as *teoyaomiqui*, a warrior woman who died in childbirth, who acts as pychopomp, a guide for the souls of the dead to their final destination. Women who died in childbirth in Aztec times were considered Goddesses, women who died fighting like a warrior, and their souls were believed to live in heaven near the Sun, accompanying him from the zenith to the sunset.

Joseph Campbell, the famous historian of religion, has suggested that the skull in her navel and genitals represent the power which death and sacrifice generates, sustains and consumes in this terrestrial plane, which corresponds to the meaning of the mask of the Sun in the so-called Calendar of the Sun. Under the colossal legs are the underworlds, the nine Aztec underworlds which culminate in "that place of duality" of feminine force, where the last transcendent "One" becomes the "Lord — Lady" of duality: Ometecuhtli, Omecihuatl, who are identical in their essence with the power of the pivotal skull.

Among the Nahua, this Earth Goddess, the mother Goddess, was known as Coatlicue, and also as Cihuacoatl "Female Serpent" or "Snake woman," who had also a warrior personality which can be seen in many of her attributes and different aspects.

On the Earth are the hills, the mountains, the rivers, the lakes, the rocks, the caves and the forests which are part of her, but at the same time have their own types of energy which makes them into receptacles of the sacred, and therefore places of worship. According to the myth, from the earth monster were formed the wells, the fountains and the caves, from her nose the valleys and the mountains, and from her hair the trees, the flowers and some nourishing plants.

It is believed by many modern ethnic groups that these places are inhabited by supernatural beings, or as the *tzotzil* say by "beings different from men," who in a sense behave in the same way as the humans.

Many a people believe that by its height, its nearness to the sky and by its inaccessibility, the mountains are a source of sacred force and that, like Mount Meru, they are the abode of supernatural beings. Among the Nahuas thc best known mythic mountain was Coatepec, the snake mountain where Huitzilopochtli, the patron God of the Mexica, was born, and which was represented by the Great Temple of Tenochtitlan, the capital city of the Mexica. Also the hills around the city were deified and many special rituals were performed on them, which were attended by the king and all the nobility. Besides in certain festivals they made replicas in a dough made of *amaranth* seed of the different mountains, and after "sacrificing" them, they were ritually eaten.

It has been suggested (López Austin and López Lujan) that the Mesoamericans believed that inside the hills there was an enormous vessel, which was open or closed by the Gods of water, so that this liquid would flow on the earth.

Even among the hills there was a hierarchy, which was dominated by the above-mentioned Popocatepetl or Don Gregorio.

We have mentioned that caves are believed to be an entrance to the underworld, and it was a place where the *graniceros* or rain makers met. Caves were also believed to be places from where the lineages of the different ethnic groups had come. For instance it was believed that from the seven caves of the famous Chicomoztoc had emerged the ethnic groups which populated the Valley of Mexico, among them the Mexica.

So important were caves that there is a version that the famous city of Teotihuacan, which flourished from the second to the seventh century, was founded around a cave in the form of a flower which is inside the pyramid of the Sun.

Plants, with their roots inside the earth may be considered as an extension of her, they grow from her as a seed and receive from her the nourishment which will make them grow, and they never abandon her.

Trees, by their size, their strength and their longevity stand out above all plants. Almost all the peoples have one or several trees which they consider sacred, the *pīpal* in India, the *baobab* in Africa and in Mexico the *ahuehuetl* in the highland and the *ceiba* among the Mayans.

Mesoamericans, as many peoples in the world, consider that these trees are the centre of the Universe and that as with mountains they serve as a means of communication between the underworld, the earth and heaven; they are also believed to be a support of heaven.

Tree symbolism seems to be especially important among the Mayas, especially as we mentioned the ceiba, which is a very high and slender and vertical tree, which stands out from the others in the tropical forest, and which has been worshipped for thousands of years by the peoples of the south of Mexico. It was a symbol of strength, used as an altar, and was used as a great roof under which many meetings of the villagers took place; but unfortunately it has been almost extinguished by the illegal woodfellers of our tropical forests.

The tree is shown very frequently in the few Mayan codexes, and in the drawings of the funerary vessels and in the tableaus of the temples of Palenque, as a cosmogram which brings together the cosmos and the vertical and horizontal spaces, as a great cross which shows the world's directions and whose roots come out from the ctonic dragon, which is the representation of Itzamna, the creator god. Through its trunk flows *itz*, the vital liquid which is equalled to sap, to blood and semen which communicates with the frond in which birds and other mythical animals stand as representations of various gods. These Mayan gods seem to be made of nature itself and one of them is the Supreme God whose epiphany is a celestial serpent playing the part of the ecliptic, or the way of the constellations. But it can also be seen as Chaak, the god of water, Bolon Dzacab, the thunder and fire, god of vegetation and of procreation.

Among the Nahuas, different types of trees pointed at the four quarters of the universe which has been represented in the Fejervary Meyer Codex and in others. It was also believed that in the paradise placed in the west there was a tree full of teats where the souls of little children sucked life from, waiting to be reborn.

Probably, the strongest link among the Mesoamerican peoples with nature is shown in their relation with maize, fundamental plant for their subsistence. It is difficult to conceive this relationship of the Mesoamerican Indian with this plant, on which their life has depended since five thousand years ago and to which they gave all the possible care which a living being deserved.

Even if around the great lakes of Tenochtitlan, there were great amounts of animal food and vegetables which could be collected, and that there were also other types of nourishing plants, like *amaranth*, beans, tomato, squash, etc., maize was still the main food, with which were prepared, besides the different types of *tortillas-chapatis*, *atole* (drink of a maize gruel), *pinole* (ground broiled maize), *tamal* (maize cakes) and in the

Mayan area *posol* (maize gruel), which is one of their most important meals; that is why they still keep considering that maize is "the body and emanation of our lord."

There are many myths which tell us how maize appeared. One of the most common is the one in which the god Quetzalcoatl, "feathered serpent," saw red ants carrying grains of maize, he converted himself into a black ant and followed them into a rock and so brought some kernels back to the other gods who made them available to men.

We have also mentioned that from the hair of the Earth monster emerged plants, including a sort of maize. In another myth it is told that a god called Piltzintecuhtli went under the Earth and from the different parts of his body, different plants came out, maize emerged from his nails.

Among different modern American Indians there is a very widespread myth in which it is told that maize was a child or a young man who sometimes is confused with Christ and with the Sun, who ends, like Christ, dying so that human beings are able to enjoy their food. In a recent myth from the *huaxteca* (Angela Ochoa : 103) it is told that the grandmother formed out of maize dough four human beings: two men and two women, "From there the flesh of men started. The men's bones were formed from the core of the maize. From there started human life. Since then men cannot live without maize. It is necessary that he (the young man) continues feeding with maize." The same people believe that a character called *Dhipak* became maize.

Among the Nahua, maize was deified with several names, the principal ones are Cinteotl, which means tender ear of corn, and Chicomecaotl is the ripe ear of corn, which are represented by the figure of a young and mature woman.

During the ritual cycle of the Mexica year of 365, divided into 18 months of 20 days, the ones dedicated to maize were the more abundant and the most important. Earth, water, sun and maize formed an inseparable quatrinomy, *without the first three there would be no maize, and without it there would be no possibility of life for men.*

Rituals took place when the plant was sown, when it was growing, when they were weeding and when it was harvested.

When the seeds were placed in the earth, they addressed them with the following words:

> Listen sister seed, you who are our nourishment, attend, princess earth, that I entrust in your hands my sister who is our nourishment . . . very soon she will come out from the earth . . . welcome her for birth . . . then very soon she will come out from earth . . . give her your good wishes for her birth, to my sister, our nourishment, Tonacacihuatl (woman of our sustenance).

Even if as I mentioned, for every stage of the cultivation of maize a ritual was made, many of which were made by the peasant in their fields, the main ceremony took place in honour of the goddess Cinteotl in the month called *hueytecuilhuitl* in autumn, when maize was ripe but still "milky."

The living image of the goddess was a young girl, who was dressed like the goddess, adorned among other things with a quetzal feather on her head, around her a series of ceremonies were performed which ended when the girl was placed on a platform in front of the goddess' sanctuary, surrounded by all kinds of fruits of the earth, vegetables, flowers, and maize cobs.

People approached her to offer their blood which was drawn from different parts of their bodies (ear, calves, etc.) and after that the young woman was sacrificed, so that her blood would sprinkle on all the fruits of the earth.

The Mayas believe that men were made from maize dough mixed with the blood of the tapir and snake.

Even today the *tzotzil* believe that maize is sacred and that it has a tangible and an intangible part, therefore it has to be treated with lots of respect and circumspection, for if it is given a careless and unrespectful treatment, the intangible aspect of its ego may get angry, be resentful and tell its pains to its spiritual guardians and ask for punishment for the ungrateful offenders.

Maize grains were and are used for divination.Among the Nahua speakers of the Puebla mountains, when they are about to plant maize, they take the kernels to a syncretic Catholic altar in their homes, where they offer candles, flower and a local incense called *copal* and they address prayers to the supernatural beings. They say the prayers in Nahuatl, but they mix them with Catholic prayers: Hail Mary and Our Father, which they use as a sort of *mantra*.

The owner of the house who acts as a ritualist blesses the maize kernels and sprinkles them with holy water.

Then he addresses the Earth, who is going to receive the seeds (Lupo 196).

> Earth of my soul
> Flat earth
> Humid earth
> Mary, Mother Trinity
> Your wife, very sacred Trinity
> Earth of my soul
> Here we are going to place
> This seed, this maize. . . .

And he goes on enumerating all the plagues and diseases which may fall to maize, and hopes to avoid through the intercession of the supernaturals, including Jesus Christ who has become for the Indians a cultural hero who besides being a representation of the Sun in some of his *avatārs* is known as Señor de los Trabajos, Master of works, who is believed to have discovered domestic plants and to be the creator of agricultural work. Some scholars (Lupo 204) think that the Lord of Works is a form of the prehispanic god and culture hero Quetzalcoatl.

The following day, after the presentation of the seed to the altar, he addresses the Lord of Works and starts the sowing, at noon he and his helpers have a meal of turkey with mole (Mexican curry) and *tortillas-chapatis*. After the meal, they make a hole in the middle of the field, and place there the bones of the *turkey* and a *torilla* divided into seven pieces as an offering to the earth.

During the time when maize plants are growing, the owner of the field has to keep sexual abstinence and abstain from some foods. There follow more rituals during the different stages of the maize growth till the harvest.

Almost all traditional Mexican Indians make offerings to help the growth of maize, offering flower, candles, copal, food and paero cut figurines which represent several deities.

We have to mention among the symbolically important plants for the Mexica the *agave maguey* and the *cactus nopal*. The first was a very useful plant, out of which everything was used; the fibres of its leaves were used for weaving, the leaves were also used for roofs, the thorns of the leaves were made into needles, the "trunk" of the flower plant was used as pillars, but what made this plant most mysterious was that from its trunk came a sweet liquid, out of which when fermented they made an intoxicating drink called *ohtli*. Very few people were allowed to drink this intoxicating drink which was considered some sort of sacred.

Mayahuel was the goddess of the plant, but there were many gods called the hundred rabbits who were related to *ohtli*. Many priests were also called this way and they were in charge of various rituals during the ceremonies which took place round the year.

While this intoxicating drink was forbidden, the brave warriors who had taken at least four captives were the people who held more prestige, used to take in their frequent dances intoxicating mushrooms and the *peyute* cactus.

Dance was a very important form of ritual and entertainment in which all the population participated, sometimes all together, and sometimes by specializations.

Nopal is a plant which grows in the dry highland. Its leaves are edible, and still today is a delicacy for the Mexicans. The fruit which is also edible has a high symbolic meaning, as the human heart which was offered to the gods is called *quauhnochtli,* the eagle cactus fruit. It is part of the Mexican foundation symbol, as the legend says that the Mexica alongwith other people started a migration which is known as a pilgrimage from their ancestral land which was somewhere in the north of Mexico into a promised land which was shown by the apparition of a sign which was an eagle standing on a cactus and eating a serpent.

We may mention in a very superficial manner the strong relationship that the mesoamericans had with animals.

They believed there was a time that they spoke the same language and that they understood each other. Some of the preceding ages — there were four — were inhabited by different animals, or when the world was destroyed the men who inhabited those era were transformed into animals.

Most of Aztec gods are anthropomorphic, though they may appear in the guise of animals too. Besides it was believed that strong men like some kings or witches could have an animal companion called *nahual* and that they could transform themselves into any animal. This belief is still extant into many Indian peoples of Mexico, and it is also believed that if you kill the *nahual* animal of the person he will also die.

Mayan deities had many animal symbols, as I had mentioned before, and most of their gods were represented in the form of animals or with animal traits.

There are three very important animal symbols among the Aztecs, the eagle and the jaguar were symbols for bravery. The eagle was the symbol of the day Sun and the jaguar was the symbol of the underworld-nocturnal Sun. And we find multiple illustrations of these two animals.

There were a sort of brotherhoods of the bravest warriors who were called the eagles or the jaguars. The brotherhood of the eagles had as their patron god the Sun and they held a special ceremony in a day of the ritual calendar called "4 movement," when they sacrificed a war prisoner as a messenger to the Sun.

The serpent is a very old and popular symbol related to earth, obscurity, darkness and fertility. The most common iconographically represented snake in Mesoamerica is the rattle snake.

From very ancient times snakes appear throughout Mesoamerican history related to earth and water, for instance in the ancient city of Teotihuacan or in Tula as pillars in the form of snakes. But the snake is most outstanding in Mayan art where it is totally polyvalent, it is related to the sky, to the *Via lactea*, to the generative principle, to earth, to water, to blood, to semen and to maize, but also to death. Which, in its cyclic and eternal movement, transforms itself into new life, and the sacred death which transforms man and links him to the gods. In this transforming aspect it appears in some lintels, when out of the mouth of the snake emerges the head of a governor who has passed through an initiatic rite.

The *jaguar* was one of the animals which caught the attention of the Mesoamerican peoples from very ancient times, because of its strength, its ferocity and many other characteristics. Unfortunately, it is on the verge of extinction as its habitat has been reduced to the remnants of forests in the South of Mexico.

There was and is also a belief that people have an animal companion, a sort of alter ego, and that powerful people are able to transform themselves into animals and that if the animal companion is hurt or dies, so does the person.

There are many beliefs about this relation with animals. To end this discourse let there be a prayer which the O'odham (Reisbaum : 60) or Papago say every year as they come to their sacred place in the northern border of Mexico:

> We perform this ceremony every year so that it rains, to purify our land, to protect living beings, to protect the people so that there is goodness in the world at all times. The earth we are watching has been terribly degraded. Wherever we see the land is drying, we see plants disappearing, we see animal and natural land disappearing. There is no water in the soil, there is no much rain, the animal species which lived in this area disappeared or are on the verge of extinction. It is the same with plants: many of the plants which were in this land are not here any more. The air which surrounds us has been degraded and is contaminated, so that it affected many things on earth. So our ceremony, through our ceremony we try to pray for all of that, we try to assure the Creator that we are doing our best efforts to preserve and protect all that has sense in our existence.

13

Life in Santal Thought

Onkar Prasad

In this paper an attempt has been made to deal with the ideas and practices related to life among the Santals, a tribe of Protoaustraloid group inhabiting mainly the eastern region of India.

Myth being an evidence of thought is presented first to know about the beginning of life. The myth of creation, in brief, runs as follows:

In the beginning there was nothing but water; under water there was earth. Then *Thakur Jiu* (the Sun-God) created the beings that live in water. Thereafter he decided to make a pair of human beings containing earth. When he had just finished making them and was going to give them life (souls), *Sin' Sadom* (the day-horse) came down from the above, trampled them to pieces and left. This made *Thakur* aggrieved. Then he left the idea of making them of earth. He decided to make two birds. He made the two *Has Hasil* (a goose and a gander) by pulling the dirt off his breast. Thereupon he breathed on them. They became alive and flew upwards. As there was no place for the birds to perch upon and food to eat they implored *Thakur* to make them available. Then *Thakur* caused the earth to be created. He also sowed sirom seed from which the sirom plant sprang first. Later he sowed the seeds of the *dhubigas*, *Karam* tree and all other vegetation to grow. The birds perched upon the sirom plant and made a nest. There they laid two eggs from which two human beings *Pilcu Haram* and *Pilcu Budhi*, the primeval couple of the Santal were born. (For detailed version of the myth see Bodding 1942).

From this myth it is revealed that Life is a creation of God and seed is the perennial source of life on earth. These two ideas are central to the Santal beliefs related to creation of life. Following the latter, the Santals have developed the physiological theory of creation. According to them there are two kinds of human seed — the male and the female, both mobile and potent. The Santals believe, without any doubt, that the male seed is responsible for the creation of life but they credit female with seed also. As outlined by Archer (1974), the Santals believe that the male seed obtains from blood marrow, which trickles down a vein to the testes. There it is made into semen and gathers in a little bag. The female seed is made in much the same way. The juice from her marrow sweeps into her womb. There it turns to semen and mounts in a little bag.

The basic rhythm in the woman's body prepares the womb for conception. The day her 'flower' ends, the man's seed enters (after copulation). It finds the flower's root from which her blood comes. It mingles with it. Then the blood gathers and the next month the flower fades (Archer 1974:158).

When conception occurs the two semens change into blood. They get bigger. The outer blood forms a bag. The ends make a cord and the bag rests in the belly. The soul of the *Jiu* is swimming in the blood. In four months a child forms and the *Jivi* settles in it. The head comes first and after it the arms, chest and legs (*ibid.* 159). The Santal believe that while it is necessary to copulate in order to get a child, however, pregnancy does not always follow intercourse. The ultimate power to grant life rests with the *Bidhi Bidhanta*. These are aspects of *Thakur Jiu* and it is they alone who mix a man and woman's seeds. A woman may hunger for a child each day but only when *Bidhan* wills it do their semens mix. If the *Bidhi Biddhanta* wills a girl, *Bidhi* gives the two sets of semen a *Jivi* and if a boy then *Bidhan* does so. A child then comes (*ibid.* 160).

However, for uniform growth of the child in the womb and for her smooth delivery the mother has to observe a number of precautions after conception of life and also after the child has been born. During pregnancy period, a woman must not look upon or touch a human corpse. She must not weep when a death occurs. In case of any difficulty an *ojha* is called upon to divinise whether a *bonga* (spirit) is present. When ascertained, a vow is taken to make the proper sacrifice to ensure speedy delivery. Midwife is engaged for delivery. When the child is born the umbilical cord is severed by the midwife using a sharp iron arrow. The placenta and the umbilical cord (after birth) are then buried near the main door of the room where the delivery had taken place. Every care is taken that they are not eaten by a dog or any other animals. Otherwise, the mother will become sick and die. The Santal also believe that if the deceased child is buried deep in the ground, the difference in age between the present child and the next will be long. If it is, however, buried near the surface, another pregnancy may be expected after a short time.

When a child is born, the house, the village and the sacred places are all polluted. Then no festival, wedding or sacrifice can take place before they are purified. A ceremony called *Met' Halan*, which literally means "lifting the eye," is observed to protect the mother and the child. In this the mother and the midwife sit facing each other on the floor of the house. The mother then fills a leaf cup with the rice-beer and gives it to the midwife who throws away the contents on her left side. This is repeated three times and is believed to protect the child and the mother from any imminent danger. The ritual cleansing ceremony after birth, called *Janam Chaṭiār,* is performed to purify the house and the whole village for giving the child a name and thus formally admitting it in one's father's clan and sub-clan, and also giving it protection of its father's spirits and incorporating the child into the tribe. (For details see Troisi 1978: 156-63).

After going through the ideas and practices associated with life among the Santal, it is understood that they consider life as the creation of the Supreme Being. Though the seed is the perennial source of life, the ultimate power to grant life rests with the Supreme Being.The birth among the Santal, like other human groups, is polluting but auspicious. Life is represented by odd number (whereas death by even number). They therefore organize their important activities following the odd number principles i.e. the life principle.

References

Archer, W.G. (1974). *The Hill of Flutes: Life, Love and Poetry in Tribal India,* New Delhi: S. Chand & Co. (Pvt.) Ltd.

Bodding, P.O. (1942). *Traditions and Institutions of the Santals*: Oslo Etnografiske Museum, Bulletin 6.

Trosi, S.J. (1978). *Tribal Religion: Religious Beliefs and Practices Among the Santal,* Delhi : Manohar.

14

Origin of Life and Piltdown Forgery

A.S. Inam Shastri and _Shaila Parveen_

God said, "Let the earth bring forth the living creature after his kind."

There are three possible explanations of the origin of life upon earth:

1. that it was supernaturally created,
2. that it was brought from some other parts of the universe, or
3. that it arose by natural processes from inorganic matter.

If we think over deeply it comes out that the first two of these explanations must be rejected because they are not, strictly speaking, explanations. Science attempts to define all that we observe in terms of natural laws; to bring in supernatural assumptions means breaking the rules of the game. It does not mean that we don't believe in existence of supernatural sphere, but if such assumptions are brought into consideration, there could be no place for science. The second explanation that life was brought from somewhere else is also unsatisfactory, because it does not explain the origin of life, but creates hindrance to trace the origin of life. So only the third explanation has to be considered to study the origin of life, i.e. life arose upon the earth from natural causes.

It appears life arose from physical agencies, acting differently on simple materials like water, carbon dioxide, and nitrogenous matters such as ammonia, when conditions were very different from those of today.

Scientific Approach

Charles Darwin, with another English man Newton, was able to present a scientific account of how evolution might have taken place: the theory of natural selection, generally described as the theory of "the survival of the fittest." Even so Darwin's work would have been impossible, if not supported by the founder of modern geology, James Hutton of Edinburgh in 1788, and other great men.

However, with the acceptance of the Darwinian theory, it comes out that man has reached his present physiological condition, passing a number of stages of evolution. Through all the ages man has wondered about the mysteries of heredity. However, it was only towards the end of the last century that a beginning was made.

The origin of life is traced on earth since uncountable years, whether it is a creation of God, or the result of natural evolution. As the life came into existence, the mystery became deep. Science is still in the way to understand the facts of life, the body, the soul, and death.

One has to believe in the scientific theories, but the fraud regarding "Piltdown Man" creates doubts to some extent. In 1953, it came to light that "Piltdown Man" was a fraud. After a long forty years of its acceptance, the matter does not seem to be resolved. As a result, some people thought that the exposure of the fraud made the scientists look foolish. Others jumped to the conclusion that it blew to bits the theory of evolution and left us no choice but to return to ancient legends about the separate creation of man out of a pile of dust. But, in fact, the removal of "Piltdown Man" from the list of human fossils does not weaken the theory of evolution. The three English Scientists: J.S. Weiner and W.E. Le Gros Clark of the University of Oxford, and Menneth P. Oakley of the British Museum detected the "Piltdown Forgery."

In 1911, Charles Danson found some old bones in a gravel bed in Piltdown. He brought them to the attention of Arthur Smith Woodward, who came to the conclusion that a primitive man with an ape-like jaw had lived in England about 5,00,000 years ago.

From the very start, scientists disagreed with each other about Woodward's conclusion. Even those who supported Woodward's views found some features of the bones of the Piltdown Man very puzzling. Finally the whole situation was thrown into confusion, when, in 1949, Dr. Oakley proved that the bones of Piltdown Man couldn't possibly be more than 5,00,000 years old.

Then, in 1953, Professor Weiner had a hunch. It occurred to him that if Piltdown Man did not seem to make sense as a fossil, perhaps it was because he was not a genuine fossil at all. Perhaps the bones of the Piltdown Man were a forgery, made up to look like a fossil and then planted in the ground. So he and his colleagues re-examined the bones. They subjected them to a long series of laboratory tests. The result of these tests unmistakably pointed to one conclusion: Professor Weiner's hunch was correct. Piltdown Man was a fraud.

The above mentioned facts indicate that forgery is possible even in scientific matters. It is not an easy matter to understand; the concept of life is a very complicated one. So, life is still a mystery which may only be well understood in the light of religious and traditional view-point. However, life has to be devoted to God for the sake of immortality.

15

Towards Understanding Life in the Context of Social Reality

Umesh K. Singh

MAN as a social being is distinct from other biological beings. He is distinct from animals as he can transcend the constraints of his environment. He does not act on instincts; rather his instincts are modified and reshaped by social and cultural conditions. In many ways the society teaches us to reject the instinctive behaviour. Whatever part instinct plays in our lives, it is complicated by cultural variations. A shared culture that specifies what we can desire and how we can attain those goals, takes the place of the biological straitjacket. In the absence of strong biological linkages, reciprocal roles provide the mechanism for coordinating human behaviour. We make life manageable by creating social institutions that do us what instincts do for animals (Bruce, 2000:24).

Human life as a dynamic process is very much determined by social and cultural settings in which one lives. We are socialized in the culture so that important elements of it become embedded in our personalities. In a society, we associate with different types of groups and institutions and follow cultural norms to survive socially and to have a stable life. Much of what we believe can be predicted from such social characteristics and repeated social patterns of gender, race, class and status.

In case of human beings, life is not merely a construct that portrays that he is a living object and has consciousness; rather there are many patterns of life and living. One talks of social life, family life, public life, religious life and so on. People have variations in lifestyle, life value, and life conditions. Meanings and interpretations assigned to one's life differ with the variations of social and cultural settings. Culture involves the lived intersubjective symbolic relationships through which we understand the conditions in which we live (Smith, 2002:83). Who we are and what we do is conditioned by the society in which we live. If there is to be any explanation of human behaviour, there must be regular patterns to life, and those regularities will at least partly be caused by forces outside our control and cognition. People create social frameworks. Some part of those frameworks may be fixed in formal law, the bulk of which is merely conventional. Thus the reality is socially constructed, our behaviour has hidden social cause, and much of social life is profoundly ironic (Bruce, 2000:18).

Life in Sociological Perspectives

Sociology is an extended commentary on the experience of daily life (Bauman, 1998:231). Through an examination of the unspoken assumptions and tacit expectations which structure our views of the world around us, sociology explores lives in society with the aim of reaching a better understanding of plurality of human desires, cravings and purposes. From Comte onwards, sociology has tried to theorize through diverse perspectives to understand life of individual in society.

Comte's Law of Human Progress has theorized three stages — theological, metaphysical, and positive. Each of these stages was characteristic of development of individual minds as well as of societies. According to Comte, the individual mind is an indirect evidence of the general mind and of human progress because each of us is aware that he is a theologian in his childhood, a metaphysician in his youth, and a natural philosopher in his manhood (Comte, 1853:27).

Spencer talks about two ideal types of society. In a militant society, individuals are coerced into various combined actions, and the will of the citizen in all transactions, private and public, is overruled by that of the government, so that cooperation is compulsory. In a militant type society, the regulating system dominates the sustaining system. In industrial societies, the sustaining system assumes more significance. Cooperation is voluntary in all relationships. State regulatory power is minimum and functions only to protect individual rights and property. The will of the citizen is supreme and the governing agent exists merely to carry out their will (Spencer, 1899:124-28).

Durkheim identifies collective consciousness, or the totality of beliefs and sentiments common to the average members of a society. This totality of beliefs and sentiments forms a determining system with a life of its own (Durkheim, 1893:39). Collective consciousness varies from one society to another because it varies according to the nature of the division of labour. Durkheim uses two conceptual terms to describe the functions of division of labour in society — mechanical solidarity and organic solidarity. Mechanical solidarity is characteristic of more primitive societies, in which division of labour is minimal and individuality is zero. The individual does not belong to himself but is literally a thing at the disposal of society. The common consciousness in this type of society is primarily religious. Religion pervades the whole of social life (Durkheim, 1893:130). The organic type of society is characterized by specialization and individuality. The collective consciousness leaves uncovered a part of individual consciousness, so that there may be established in it those special functions that it cannot regulate (Durkheim, 1893:85). Durkheim's emphasis is on social facts which he defines as "ways of acting, thinking, and feeling, *external* to the individual and endowed with the power of *coercion*, by reason of which they control him" (Durkheim, 1895:3).

Marx, like Hegel, perceives *society's* reality as imperfect, even negative (Marcuse, 1941:312). This imperfection arises in the historical process: "All history is nothing but a continual transformation of human nature" (Marx, in Ollman, 1971:81). The transformation involves the alienating of person, so that the human as social being is replaced by the individualist and egoist (Marx, 1844:138).

A central theme in Max Weber's work is his concern with the problems of Western civilization, especially the rationalization and demystification of all aspects of modern social life — "the disenchantment of the world" (Weber, 1946:155). He is concerned with the radical transformations in social life that distinguished modern from traditional society. He emphasizes on understanding the meaning, causes, and consequences of social action. Weber identifies four types of meaningful social action (Weber, 1925, 1:25-26) — instrumental rational action, value-rational action, affectual action, and traditional action.

Both Marx and Weber investigate capitalism, focusing on the predicament of modern individuals when the economy becomes human destiny (Lowith, 1993:48). And both theorists take a pessimistic view: for Marx, it is alienation; for Weber, rationalization. Weber finds most troubling the way that rationalization produces a "disenchantment of the world" involving a retreat of the "ultimate and most sublime values"from public life into private life, with the resulting loss of any sense that the world has meaning (Weber, 1919:155).

In *The Protestant Ethic and The Spirit of Capitalism*, Weber illustrates how the "specific and peculiar rationalism of Western culture," tied to the "technical and economic conditions of machine production" is originally motivated by the religious values. But in time religious values fall by the wayside, and the technical and economic aspects became an irresistible force determining everyone's lives and enclosing them like an "iron cage" (Weber, 1904-1905:181).

In Simmel's analysis, the relationship between individual and society is dialectical. The individual "exists both for society and for himself" so that the individual is "contained in sociation and, at the same time, . . . confronted by it" (Simmel, 1908:17). Individuals are born into and influenced by their social and cultural milieu, but as self-conscious individuals they stand opposed to that milieu, and out of that opposition they transform the milieu. According to Simmel, the forms of social life as well as the individual personality are characterized by this ambivalence. Conflict, for example, is an integral and inevitable part of social interaction, as is cooperation; a relationship or interaction totally devoid of conflict is logically, and empirically, impossible. He suggests that "happiness, though it is the object of all endeavours, would be more boredom if it were ever achieved as an eternal state"(Simmel, 1900:166).

Parsons is a "grand-scale" theorist. Parsons states that action is "behaviour-oriented to the attainment of ends in situation, by means of the normatively regulated expenditure of energy" (Parsons and Shils, 1951:53). Modes of orientation or motivation for individual action include the cognitive, cathectic, and evaluative. These terms are close to the everyday meaning of thinking, feeling and valuing or willing.

Looking at individual choices from a macro-cultural perspective, Parsons and Shils (1951:172-83) state that different cultures guide an individual toward one or the other set of dichotomous choices. Parsons notes that one set of choices is dominant in the traditional society, while the other usually occurs in the modern society. These dichotomous choices, which he calls the pattern variable, are: affectivity/affective neutrality, collectivity/self, particularistic/universalistic, ascription/achievement, and specificity/diffuseness.

At the micro level, the interactionist, phenomenological and ethnomethodological approaches place a strong emphasis upon gaining entry into and having direct access to lives of those being studied (Smith, 2002:31). These interpretive sociologies embrace the subjective dimension of social life wholeheartedly in the research process and regard the existence of object of analysis as the product of our interpretations rather than treating such things as real. In addition, they focus on the shared cultural practices through which identities are constructed and engage in an exploration of self, interaction and meanings.

Interactionism treats social actors and their small face-to-face interactions as the basis of all social life. It suggests that the meaning of any concept or idea (personal, political, philosophical or scientific) could only be located in the experiential conditions in which it is produced (Blumer, 1969:2). The self is not a rational calculating machine but is a changing product of an ongoing process of interaction between individuals and the meaningful interpretations of unpredictable relationships; in short everyday life involves informed guesswork.

In the phenomenological approach, consciousness and the production of meaning are portrayed as the result of the intersubjective relations between actors engaged in mutual discovery. Alfred Schütz, in *The Phenomenology of the Social World* (1932), argues that it is the condition of intersubjectivity through which the individual actors are able to grasp each other's consciousness and construct their life world. By sharing time and space, individual actors can engage in a process of understanding which involves the discovery of what is going on in the other person's mind (Schütz, 1932:112-13).

Ethnomethodology, pioneered by Harold Garfinkel (1967), combines Schütz's emphasis on the 'taken-for-granted' common-sense assumptions (tacit knowledge) of everyday life with the empirical research strategies of anthropologists and symbolic interactionists. Ethnomethodology seeks to find ways of locating and reconstructing the taken-for-granted common-sense assumptions of social practices upon which interaction rests. In the ethnomethodological approach, human actors are constantly attempting to make sense of mass of sensations they experience by drawing upon their stock of stories and meaningful interpretations. Ethnomethodology sees the objective reality of social facts as an "ongoing accomplishment of the concerted activities of everyday life" (Garfinkel, 1967:vii).

Sociological perspectives and theories have enlarged further in the present discourses on modernity and postmodernity. The new focus is more on the meaning of life in the changing scenario of the postmodern society rather than on the development of a narrow theoretical paradigm following any of the stagnant conceptual framework.

Context of the Modern and the Postmodern Conditions

Man's life has been constantly on change with the change in his existential condition primarily because of innovation of technology and change in social and economic structure of society. From hunting and gathering stage to horticulture and agrarian and then to industrial stage, man's life and way of living has always been getting reshaped and reformed.

In modern society, the Enlightenment has brought about a social order characterized by scientific knowledge and rationality. Modernity has replaced the set of rules given by tradition, questioning all conventional ways of doing things, substituting authorities of its own, based in science, economic growth, democracy or law. The transformation which began in Europe has altered the world in unprecedented and irreversible ways. Because of technological developments, the routines of everyday life are altered, for instance, when we no longer have to rely on face-to-face relationships in order to communicate. Our social relations become stretched over time and space, connected by TV signals and fibre-optic cables. More and more, we do things at a distance. (Lyon, 2002:28). Differentiation has spread to all social spheres. Not only is "work" split off from "home" but also from "leisure," "religion" and so on. "Public" life is by the same token distinguished from "private" in a novel way, and alongwith these the lives of men and women have also been redefined in terms of specialist tasks. Rationalization, that is gradual adoption of a calculating attitude towards more and more aspects of life, has become a central motif of our modern life. To observe, to calculate, these are the hallmarks of modernity. The rational approach that underlay science and that found its most dynamic expression in the capitalist economy has taken centre stage, systematically entering every sector of society. A further sphere of rational organization was the city. Punctuality, calculability, exactness are forced upon life by the complexity and extension of metropolitan existence. Here in microcosm one would see differentiation, commodification and rationalization writ large. Modern life is marked more by formal and contractual relations. People are distancing themselves from overly intimate relationship. The society of strangers has appeared and flourished in the city. Older way of keeping order such as the public and brutal treatment of offenders, gave way to modes that emphasized self-discipline, self-control. Such discipline came to characterize many social contexts.

Differentiation continues to divide social life into increasingly self-contained spheres. The rule for each sphere — work, leisure, family, citizenship and so on — unlike the fairly unified rules and conventions of tradition, are developed within that sphere alone. The modern differentiation introduces decisively self-referential systems, which from an individual's perspective have constantly to be negotiated. Far from the single, over-arching world of everyday life experienced in traditional societies, modernity brings a "pluralization of lifeworlds." The condition of anomie is reproduced again and again as 'homeless minds' accompany modernization (Berger, 1974).

Ironically, while modern liberal society is supposed to free people for involvement in more diverse ends, these actually fade into obscurity as people become enslaved to supposedly neutral techniques and technologies. While urban space is clean and orderly, it is socially and spiritually dead, except where life is sustained, paradoxically by noise, congestion and chaos. Because of increase in ethnic and cultural variety in great cities, larger portions of population are difficult to classify. Neither friends nor enemies, these are both neighbours and strangers; unclassifiable others. "Strangers often seem to be suspended in the empty space between a tradition which they have already left and a mode of life which stubbornly denies them right of entry" (Sarup, 1996:10). Such strangers tend to be stigmatized, as a reminder of their difference.

Despite appearances, modernity has lived with inner doubts and contradictions from the start. And these are not just abstract or amorphous "cultural" phenomena; they have been bound up with the messy materiality of money, machines, and streets. Modernity has bequeathed to us a world split into social segments each governed by its own rules, implicit and explicit. Authority supposedly shifts from religious to scientific grounds, but in fact the main rule of thumb is instrumental, pragmatic: does it work? Is it efficient? The autonomous self moves to centre stage, claiming new liberties that will be enshrined in civil, political, social rights. But simultaneously that individual self loses a sense of meaning and purpose, a situation that has become a fundamental problem.

A variety of major social and cultural changes is taking place within many societies. In the social condition of the twenty-first century, the key carriers of the post-modern — new technologies and consumerism — has emerged in thoroughly modern context, but so profoundly affect everyday social life that modernity itself is changing shape. We have a new society-in-the-making which best captures today's transformation or modernity. We have a society where consumer lifestyles and mass consumption dominates the waking lives of its members. Rise of media, especially television, and consumerism has created virtual reality in our everyday life. To many people much in their life is anything but simulation. We live in a *society of the image*, primarily concerned with the production and consumption of mere "simulacra" (Butler, 2002:112). In a knowledge-dominated technologically driven society, information is one of the main things we buy. In an incredibly driven, information-soaked world, there is too much of everything. Be it in magazines and advertisements or on the TV, we are subjected everywhere to a sensory overload of images. Increasingly central to the analysis of postmodern life-condition is the process of globalization. Facilitated by media of electronic information and communication, the globalization of contemporary social life is creating new situations. Not only the economic and cultural globalization has made our life more consumeristic and influenced by the world created by media, the conditions of globalization have also created a crisis of identity and integration, especially in a traditional society like India. The post-modern world is also characterized by fragmentation. The rational exists alongside the irrational, the modern pervades along the traditional, the present is pervasive along the past. There is hybridity in every sphere of life. The social reality has become a hydraheaded monster. It is grotesque as well as sublime. To face up to the implications of a fractured modernity and today's globalizing, technologized and consumerist social realities, the idea of an appropriate way of living and balance between life-spheres is pervasive and important for human beings — a way that applies to both personal and collective life. The promotion of mutual understanding and tolerance as paramount condition of shared freedom can make human life and human cohabitation more meaningful in the contemporary society.

References

Bauman, Z. (1998). *Thinking Sociologically*. Oxford: Blackwell Publishers Ltd.

Berger, P. (1974). *The Homeless Mind*. Harmondsworth: Penguin.

Blumer, H. (1969). *Symbolic Interactionism: Perspective and Method*. Englewood Cliffs, NJ: Prentice Hall.

Bruce, S. (2000). *Sociology—A Very Short Introduction*. New York: Oxford University Press Inc.

Butler, C. (2002). *Postmodernism—A Very Short Introduction*. Oxford: Oxford University Press.

Comte, A. (1853). *The Positive Philosophy of Auguste Comte* (freely translated and condensed by Harriet Martineau). New York: Calvin Blanchard.

Durkheim, E. (1893/1984). *The Division of Labour in Society* (W.D. Halls, Trans.). London: Macmillan.

Garfinkel, H. (1967). *Studies in Ethnomethodology*. Englewood Cliffs, NJ: Prentice Hall.

Lowith, K. (1993). *Max Weber and Karl Marx*. London: Routledge.

Lyon, D. (2002). *Postmodernity, 2nd edition*. New Delhi: Viva Books Private Limited.

Marcuse, H. (1941). *Reason and Revolution*. New York: Oxford Press.

Marx, K. (1844/1964). *Economic and Philosophic Manuscripts of 1844*, New York: International Publishers.

Ollman, B. (1971). *Alienation: Marx's Conception of Man in Capitalist Society*. Cambridge: Cambridge University Press.

Parsons, T. and Shils, E. (Eds.) (1951). *Toward a General Theory of Action*. Cambridge, MA: Harvard University Press.

Sarup, M. (1996). *Identity, Culture, and the Postmodern World*. Edinburgh: Edinburgh University Press.

Schütz, A. (1932/1967). *The Phenomenology of the Social World*. Evanston, IL: North Western University Press.

Simmel, G. (1900/1978). *The Philosophy of Money* (Tom Bottomore and David Frisby, Trans.). London: Routledge & Kegan Paul.

Simmel, G. (1908/1971). "How is society possible"? In Donald N. Levine (Ed.), *On Individuality and Social Forms* (pp. 6-22). Chicago: University of Chicago Press.

Smith, M.J. (2002). *Culture—Reinventing the Social Sciences*. New Delhi: Viva Books Private Limited.

Spencer, H. (1899/1969). *The Principles of Sociology* (3 Vols.) (Stanislav Andreski, Ed.). New York, Macmillan.

Weber, M. (1904-1905/1958). *The Protestant Ethic and the Spirit of Capitalism* (Talcott Parsons). New York: Charles Scribner's Sons.

Weber, M. (1919/1946). "Science as a Vocation." In H.H. Gerth and C. Wright Mills, *From Max Weber: Essays in Sociology* (pp. 129-56). New York: Oxford University Press.

Weber, M. (1925/1978). *Economy and Society* (2 vols.) (Guenther Roth and Claus Wittich, Eds.). Berkeley: University of California Press.

Weber, M. (1946). *From Max Weber: Essays in Sociology* (H.H. Gerth and C. Wright Mills, Trans. and Eds.). New York: Oxford University Press.

List of Contributors

Ajeya Jha, is a Scientist associated with the Manipal Institute of Technology, Sikkim.

Anthony Sudil, IMS, is a Professor of Christian Theology, involved in inter-religious dialogue, Christ Nagar, Varanasi.

A.S. Inam Shastri, is a Professor of Social Work at the Mahatma Gandhi Kashi Vidyapeeth, Varanasi.

B.D. Tripathi, is an eminent Sociologist and a former Professor of Mahatma Gandhi Kashi Vidyapeeth, Varanasi.

Baidyanath Saraswati, is an Anthropologist and Founder of the N.K. Bose Memorial Foundation, Varanasi.

Jogananda Jha, is a Physician associated with the Maharaja Hospital, Darbhanga, Bihar.

K.N. Sahay, has been a Professor of Anthropology at the Ranchi University, Ranchi.

Onkar Prasad, is a Professor of Anthropology at the Visva Bharati University Santiniketan.

Pratul Kumar Saraswati, is a Professor in Earth Sciences at the Indian Institute of Technology, Mumbai.

R.C. Sharma, is an eminent Indologist and Director of Jnana Pravaha: Centre for Cultural Studies, Varanasi.

Surendra Nath Mishra, is a Scientist associated with Agricultural University, Mandsaur, Madhya Pradesh.

Shaila Parveen, is a Teacher in Social Work at the Mahatma Gandhi Kashi Vidyapeetha, Varanasi.

Uma Marina Vesci, came to India in 1963 to begin a new life that few scholars have the courage to undertake, to ponder the words of ancient Sanskrit.

Umesh K. Singh, is a Sociologist associated with Kanpur University.

Yolotl González Torres, a renowned Anthropologist, Director of the Department of Ethnology and Social Anthropology, Mexico.

Index